MINTUS – Beiträge zur mathematisch-naturwissenschaftlichen Bildung

Series Editors
Ingo Witzke, Mathematikdidaktik, Universität Siegen, Siegen, Germany
Oliver Schwarz, Didaktik der Physik, Universität Siegen, Siegen,
Nordrhein-Westfalen, Germany

MINTUS ist ein Forschungsverbund der **MINT**-Didaktiken an der Universität Siegen. Ein besonderes Merkmal für diesen Verbund ist, dass die Zusammenarbeit der beteiligten Fachdidaktiken gefördert werden soll. Vorrangiges Ziel ist es, gemeinsame Projekte und Perspektiven zum Forschen und auf das Lehren und Lernen im MINT-Bereich zu entwickeln.

Ein Ausdruck dieser Zusammenarbeit ist die gemeinsam herausgegebene Schriftenreihe *MINTUS – Beiträge zur mathematisch-naturwissenschaftlichen Bildung*. Diese ermöglicht Nachwuchswissenschaftlerinnen und Nachwuchswissenschaftlern, genauso wie etablierten Forscherinnen und Forschern, ihre wissenschaftlichen Ergebnisse der Fachcommunity vorzustellen und zur Diskussion zu stellen. Sie profitiert dabei von dem weiten methodischen und inhaltlichen Spektrum, das MINTUS zugrunde liegt, sowie den vielfältigen fachspezifischen wie fächerverbindenden Perspektiven der beteiligten Fachdidaktiken auf den gemeinsamen Forschungsgegenstand: die mathematisch-naturwissenschaftliche Bildung.

More information about this series at https://link.springer.com/bookseries/16267

Frederik Dilling · Simon F. Kraus
Editors

Comparison of Mathematics and Physics Education II

Examples of Interdisciplinary Teaching at School

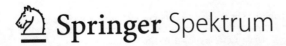

Springer Spektrum

Editors
Frederik Dilling
Mathematics Education
University of Siegen
Siegen, Germany

Simon F. Kraus
Physics Education
University of Siegen
Siegen, Germany

ISSN 2661-8060   ISSN 2661-8079   (electronic)
MINTUS – Beiträge zur mathematisch-naturwissenschaftlichen Bildung
ISBN 978-3-658-36414-4   ISBN 978-3-658-36415-1   (eBook)
https://doi.org/10.1007/978-3-658-36415-1

Responsible Editor: Marija Kojic
This Springer Spektrum imprint is published by the registered company Springer Fachmedien Wiesbaden GmbH part of Springer Nature.
The registered company address is: Abraham-Lincoln-Str. 46, 65189 Wiesbaden, Germany

Contents

List of Contributors

Le Tuan Anh Faculty of Mathematics and Informatics, Hanoi National University of Education, Hanoi, Vietnam

Nguyen Van Bien Faculty of Physics, Hanoi National University of Education, Hanoi, Vietnam

Tran Ngoc Chat Faculty of Physics, Hanoi National University of Education, Hanoi, Vietnam

Nguyen Phuong Chi Faculty of Mathematics and Informatics, Hanoi National University of Education, Hanoi, Vietnam

Frederik Dilling Mathematics Education, University of Siegen, Siegen, Germany

Jochen Geppert Mathematics Education, University of Siegen, Siegen, Germany

Daniela Götze Mathematics Education, University of Siegen, Siegen, Germany

Tuong Duy Hai Faculty of Physics, Hanoi National University of Education, Hanoi, Vietnam

Sascha Hohmann Physics Education, IPN—Leibniz Institute for Science and Mathematics Education, Kiel, Germany

Kathrin Holten Mathematics Education, University of Siegen, Siegen, Germany

Simon F. Kraus Physics Education, University of Siegen, Siegen, Germany

Vu Dinh Phuong Faculty of Mathematics and Informatics, Hanoi National University of Education, Hanoi, Vietnam

Felicitas Pielsticker Mathematics Education, University of Siegen, Siegen, Germany

Philipp Raack Physics Education, University of Siegen, Siegen, Germany

Gero Stoffels Mathematics Education, University of Siegen, Siegen, Germany

Ina Stricker Physics Education, University of Siegen, Siegen, Germany

Chu Cam Tho Research Division On Educational Assessment (RDEA), The Viet Nam Institute of Educational Sciences (VNIES), Hanoi, Vietnam

Ingo Witzke Mathematics Education, University of Siegen, Siegen, Germany

Introductory Remarks and Concept of the Book

1

Frederik Dilling and Simon F. Kraus

The presented volume is the outcome of our efforts to enhance and further develop cooperation and cross-linking between mathematics and physics education. Like the first volume (Kraus & Krause, 2020), this book is based on a collaborative project between the Hanoi National University of Education in Vietnam and the University of Siegen in Germany. The Inter TeTra (Interdisciplinary Teacher Training) project aims to contribute to the development of teacher training in Vietnam by promoting the integration and interweaving of content from mathematics and physics at theoretical and practical levels. In each case, educational approaches and concepts should be given greater consideration in order to foster optimal learning outcomes and advance the development of competencies. The main outcome of the Inter TeTra project is the conception of a university module, consisting of a lecture, a seminar, and in-service teacher training. The project is supported by the German Academic Exchange Service (DAAD) with funds from the Ministry for Economic Cooperation and Development (see Fig. 1.1).

While the first volume was dedicated to a comparison of the theoretical foundations of the aforementioned subjects, this volume will focus on practical implementation in the classroom. With this division into theoretical and practical volumes, the books reflect the conception of the Inter TeTra project (Kraus et al., 2018; Krause

F. Dilling
Mathematics Education, University of Siegen, Siegen, Germany
e-mail: dilling@mathematik.uni-siegen.de

S. F. Kraus (⊠)
Physics Education, University of Siegen, Siegen, Germany
e-mail: kraus@physik.uni-siegen.de

Fig. 1.1 This book was written as part of the Inter TeTra project, which is funded by the German Academic Exchange Service with funds from the Ministry for Economic Cooperation and Development

DAAD — Deutscher Akademischer Austauschdienst / German Academic Exchange Service

Federal Ministry for Economic Cooperation and Development

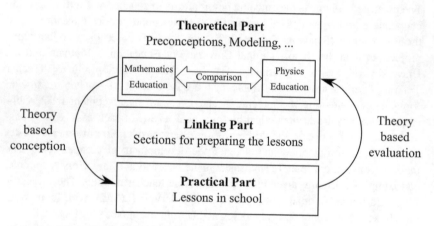

Fig. 1.2 The structure of the underlying courses relies on the interconnection of theoretical considerations and practical applications

et al., 2019). The courses also start with a theoretical comparison, based on which concrete teaching proposals are developed (Fig. 1.2). The theory-based development as well as the final evaluation and reflection on the results of the implementation represent an essential design element of the courses of the Inter TeTra project.

As a transition between the theory-oriented first volume and the practice-oriented volume realized here, the connection is exemplified in two chapters. The first chapter[1] shows how a teaching sequence on the lens equation in geometrical

---

[1] See Chapter "Educational use of Ludwig's methodology using the example of the lens equation".

optics can be designed and implemented in the style of Günther Ludwig's methodology. The second chapter[2] demonstrates the inherent importance of equations in physics and thus in physics education. In both chapters, the close connection between mathematics and physics and how they can be applied in the classroom are described in detail.

Afterwards, educational perspectives and concepts related to specific teaching topics are compared and contrasted. The objective is to identify different perspectives and their effects on teaching. Our understanding of interdisciplinary cooperation does not aim to develop uniform terms and concepts, since this would contradict the character of the two subjects, which are closely linked but are and should remain fundamentally independent disciplines. Therefore, in our opinion, a unification of the didactic principles is also incompatible with an adequate representation of the nature of the subjects (in the sense of the Nature of Science and the Beliefs of Mathematics) in the classroom and at universities.

Based on the aforementioned comparisons, exemplary lesson plans are developed, which show how a context-oriented implementation of these topics including authentic problems can take place in the classroom. The lesson plans are often based on local and regional examples, such as architecture, and are therefore particularly motivating for learners from that particular region, though also interesting in international and intercultural contexts. However, they can be adapted to the specifics of other regions and cultures with little effort.

The selection of the individual subject areas for this book is based on a comparative analysis of mathematics and physics textbooks (Dilling, 2019). Several topics have been identified that are explicitly or implicitly covered in both subjects and within which a variety of intersections can be identified. These topics are clustered as follows:

1. Numbers, Quantities, and Units
2. Equations
3. Functions
4. Vectors
5. Derivations and Integrals
6. Probabilities
7. Straight Lines and Conic Sections

---

[2] See Chapter "Equations as a tool for hypothesis formulation in physics".

Due to the breadth of the individual subject areas, content had to be prioritized since an educational comparison often cannot cover the entire field in adequate detail. The same applies to the lesson plans, in which concrete individual or double lessons within the spectrum of the respective topic are presented.

The topics of "Numbers, Quantities, and Units" are very broad and include essential basic concepts in both mathematics and physics. Possible links include measurement, mass, time, length, area, volume, and temperature. The educational comparison highlights measurement (i.e., the quantitative recording and representation of quantities). In the corresponding example lesson, the units of centimeters and meters are developed in a student-centered elementary school double lesson and linked with anchor examples, which enable the learners to deal intuitively with measurements at these orders of magnitude.

The subject area "Equations" is an integral part of both mathematics and physics teaching, covered in many lessons in both disciplines. Advanced mathematics lessons, as well as quantitative physics lessons, are hardly imaginable without the use of formulas and equations. The proposed lesson presents a possible application of trigonometric equations using a water wheel as a starting point. Physics and mathematical approaches are combined here: students determine the dimensions of the water wheel and, based on this, investigate the rotation of the wheel over time and the associated potential and kinetic energies. Here, the close connection between the topic and functional relations becomes apparent; the comparison is made explicit in the conclusion of the lesson.

"Functions" also represents a wide range of topics, including various types of functions and forms of representation. There are concrete links between trigonometric functions and harmonic oscillation, linear/quadratic functions and kinematics, and proportional relationships and Hooke's law. In advanced classes, applications for exponential and logarithmic functions can be found in radioactive decay and the perception of light, sound, and heat. Further intersections arise in the case of chain lines or the barometric height formula (i.e., the function of the decrease of air pressure with increasing height). The educational comparison primarily focuses on various representation methods, which in mathematics include verbal, numerical, graphical, and symbolic representations.

From a physics point of view, the focus is often on converting tabular measured values to graphical representations. Idealizations can quickly lead to representations that do not correspond with physical reality. The topic of functions is presented in a practical form in four lesson plans. The first lesson plan is dedicated to quadratic equations, which are worked out based on architectural and other practical examples. This is followed by the determination of the maximum throwing distance of an object as a function of the throwing angle, which is relevant to everyday life. Here, the mathematical deduction and the subsequent experimental verification are intertwined. Another lesson deals with temperature-dependent resistance and the resulting functional relationships, whose mathematical correlations allow one to make statements about the underlying physical processes. The final topic is harmonic oscillations, which are essential for the description of periodic processes in physics. Here, the functional relationships learned in mathematics lessons are applied to describe periodic processes.

"Vectors" are another topic that is important in both subjects but often presented in different ways. Vectors can be conceptually described in three ways: via arrow classes, as n-tuples, and through vector spaces. Students' first contact with vectors, however, often takes place in physics classes, where a different view prevails. Common applications in physics include forces, velocities, translational motion and, at advanced levels, vector fields. The two lesson plans presented in this book are designed for tenth-grade students and first address the basic concept of vectors as arrows and then the decomposition and addition of forces through various examples.

Of great importance in the higher grades are "Derivatives and Integrals." Physics also provides many applications in which it is impossible to determine an exact solution without the use of infinitesimal calculus. Besides dynamics, these include the fields of alternating currents, electromagnetic induction, and many nontrivial cases of mechanical work. In the educational comparison, it is shown how students can achieve better conceptual understanding by applying their knowledge of integral calculus. The practical examples demonstrate one application concerning the calculation of forces and another related to the charging and discharging of capacitors.

A topic that has found its way into physics relatively late in its historical development is "Probabilities." In addition to radioactivity, quantum physics is beginning to appear in the classroom, and neither topic can be treated without the elementary concept of probability. Probability theory is also present in mathematics lessons via game theory, which allows for a cooperative approach. The topic of statistics is also found in both subjects to a certain degree and physics cannot be imagined without it (in terms of the treatment of statistical errors or fundamental phenomena such as Brownian motion). The comparison here focuses on stochastics with an emphasis on probability theory (which the mathematician David Hilbert has called a physical science). The applications in physics are correspondingly numerous, especially in modern physics. Complementary to this, the lesson plan represents the use of statistical methods in determining gravitational acceleration. Without these methods, the empirical testing of physical laws in experiments cannot be successful.

Furthermore, in the field of geometry, there are many possible connections, especially regarding "Straight Lines and Conic Sections." At a basic level, these connections include the intercept theorems. In physics, conic sections also appear in the form of elliptical orbits in Kepler's laws and the treatment of the moon's orbit. The comparative chapter takes up the intercept theorems as a common topic and presents various practical examples. In addition to the educational comparison, the plan also presents a lesson in which various approaches to conic sections in nature, architecture, and technology are adopted and applied to teaching.

The objective of this volume is to elaborate on and compare the different perspectives of mathematics and physics on their shared topics. This should raise awareness and ensure that these different perspectives are given appropriate consideration in the classroom. The practical examples are intended to show how to develop lessons that are reality- and application-oriented, encourage active learning, and incorporate modern media so that learners are not only motivated but also develop a deeper understanding of the underlying concepts of mathematics and physics classes.

References

Dilling, F., Holten, K., & Krause, E. (2019). Explikation relevanter Inhalte für den interdisziplinären Austausch zwischen Mathematik- und Physikdidaktik. *Mathematica Didactica, 42*, 1–18.

Kraus, S., & Krause, E. (Hrsg.). (2020). *Comparison of mathematics and physics education I. Theoretical foundation for interdisciplinary collaboration.* Springer (MINTUS – Beiträge zur mathematisch-naturwissenschaftlichen Bildung).

Kraus, S., Krause, E., & Dilling, F. (2018). Inter-TeTra – A German-Vietnamese project combining physics and mathematics didactics. *Vietnam Journal of Education, 5*, 1–8.

Krause, E., Kraus, S., Geppert, J., Dilling, F., Holten, K., Nguyen, C. P., T. Chat, T. Le, C. Tho, & Nguyen, B. V. (2019). Inter TeTra—Description of a project partnership between Siegen (Germany) and National University of Education. In U. T. Jankvist, M. van den Heuvel-Panhuizen, & M. Veldhuis (Chairs), *Proceedings of the Eleventh Congress of the European Society for Research in Mathematics Education (CERME11). Symposium conducted at the meeting of Freudenthal Group & Freudenthal Institute, Utrecht University and ERME.*

Krause, E., Dilling, F., Kraus, S., Chi, N. P., Chat, T. N., & Bien, N. V. (2020). Relevant content for a scientific collaboration in mathematics and physics education research: A comparative content analysis of handbooks and conference proceedings in Germany and Vietnam. *EURASIA Journal of Mathematics, Science and Technology Education.*

Historical Relations of Mathematics and Physics—an Overview and Implications for Teaching

Frederik Dilling and Simon F. Kraus

2.1 The Development of the Modern Sciences

If one studies the history of the modern sciences, it becomes clear that they were inextricably linked for a long period of time. The division into individual disciplines took place relatively late and further development was still characterized by mutual influence. If we look at physics as an example, it can be noted that humans have always dealt with the physical world. For a long time, however, this was done only by means of observation and in a philosophical way, which is why it was referred to as natural philosophy. At the same time, the ancient Greek natural philosophers were already using mathematics to describe nature. For the Pythagoreans,[1] the natural numbers were the reference point by which they oriented themselves. For example, the distances between the Earth and the Moon and the other planets were supposed to correspond to certain numerical ratios that were considered harmonious. Similar attempts can also be found in the early

---

[1] Pythagoreans are the followers of the teachings of Pythagoras of Samos (ca. 570–510 BCE), to whom significant contributions to mathematics are attributed.

F. Dilling (✉)
Mathematics Education, University of Siegen, Siegen, Germany
e-mail: dilling@mathematik.uni-siegen.de

S. F. Kraus
Physics Education, University of Siegen, Siegen, Germany
e-mail: kraus@physik.uni-siegen.de

F. Dilling and S. F. Kraus (eds.), *Comparison of Mathematics and Physics Education II*, MINTUS – Beiträge zur mathematisch-naturwissenschaftlichen Bildung, https://doi.org/10.1007/978-3-658-36415-1_2

work of Johannes Kepler (1571–1630) to describe the structure of the solar system (cf. Galili, 2018). Of course, this is more a kind of numerical mysticism than a real mathematical approach.

Only after Galileo Galilei combined theoretical considerations, mathematical descriptions and systematic experimental investigation have we used the term physics in the modern sense.[2] However, the conceptual separation of the sciences was by no means accompanied by a clear separation between the actors in the two subjects. Thus, in the case of many historical personalities (e.g., Newton, Euler, Lagrange, Fourier, or the Bernoullis), it is not obvious whether they should be categorized as mathematicians or physicists (Uhden, 2012). This chapter attempts to show where the common roots of physics and mathematics lie and what pedagogical conclusions can be drawn from them. Since the interconnections are incredibly rich, however, we will limit ourselves to those subject areas that will be addressed later in the book, in the chapters comparing educational methods.

2.2   Scientific Problems as a Driving Force of Mathematics

One of the oldest branches of mathematics is geometry. Even from the name, which originates from the Greek for "earth measurement," it is clear that there is a close connection between mathematics and the physical world (Hischer, 2012, p. 1). It can also be shown for other fields of mathematics that they developed from the problem of concrete application. The statement that the driving force of mathematics has always been to solve problems can be traced back to David Hilbert (Kjeldsen, 2015). For the further development of mathematics, the formation of mathematical concepts was essential; however, they often do not originate from mathematics itself but represent external influences that can frequently be traced back to physics. If one follows individual concepts back to their origins, it also becomes clear why the mathematical description of nature is so successful. This phenomenon should by no means come as a surprise, as it follows directly from the historical development of mathematics (Kjeldsen, 2015). It should also be noted that only a fraction of the available mathematical concepts are used in physics at all, so that their successful application may seem even less mystical (Galili, 2018).

---

[2] See also Chapter "The Mathematization of Physics Throughout History" in Volume 1 (Tran et al., 2020).

The increasing mathematization of physics, however, also brought criticism. Representatives of the discipline saw themselves increasingly excluded from participating in its further development in the post-Newton era, which was also due to the fact that there was soon no room for an intuitive approach:

> The counter-intuitive effects of the mathematization of physical phenomena only began to be perceived with the development of dynamics, that is, the mathematization of the concept of force, as the cause of change in the state of motion. (Gingras, 2001)

Even in modern times, such turnarounds still took place, as seen in the example of Nobel Prize winner Johannes Stark, who after 1913 increasingly turned away from Planck's quantum hypothesis, of which he was initially one of the earliest supporters. Historians of science attribute this to the increasing mathematization of physics and not to the unusual new physics that disturbed Stark's experimental work[3] (Metzler, 2019).

2.3   Historical examples and educational remarks

The following examples of the common problem-oriented foundations of mathematics and physics were selected based on the school-relevant subjects in this volume. Here we want to give a basic overview of the historical development of the disciplines and offer some preliminary educational conclusions. A comprehensive analysis from a historical point of view should be left to special literature. A detailed didactic discussion from a modern point of view will be provided in Part B of this volume.

2.3.1   Numbers, Quantities, and Units

Counting objects and quantifying certain properties are basic practices that have been used by all cultures since the earliest times (Himbert, 2009). Practically inseparable is the concept of measurement, the objective of which is "the expression of characteristics of systems in terms of numbers" (Himbert, 2009, p. 25).

---

[3] Stark finally turned to so-called German physics, which was close to National Socialism, and is therefore one of the most controversial figures in physics in Germany (Hoffmann & Walker, 2006).

The earliest measurements were also based on pure counting in the form of whole numbers that referred to a certain unit. Then, there was the handling of ratios of these whole numbers and corresponding quantities. The extension of the rational numbers to the real numbers did not take place until the nineteenth century (cf. Jahnke, 2003).

As far as the reference units themselves are concerned, they have only been gradually standardized worldwide since the introduction of the SI units (e.g., the meter and the kilogram in 1889). In the process, an increasing degree of abstraction can be observed. For instance, a meter was originally defined in France in 1790 as one ten-millionth of the Earth's meridian quadrant (the distance between the equator and the pole along a meridian arc). A decoupling from this natural measure then occurred in 1889 with the introduction of the primordial meter, intended to make the base unit independent of the inaccuracies of measurement and take into account that different meridian arcs have different lengths due to the irregular shape of the Earth. In the definition valid since 2019, all SI units are now defined by physical constants (according to the current theory), which means that the meter is indirectly defined based on the distance light travels in a vacuum within a second.

However, for the determination of physical laws, as the relationship of various physical quantities, the available instruments were hardly suitable for most of history. An exception to this is classical astronomy, which, through the simple magnification of its instruments, achieved highly accurate angle measurements and precise models of the world in the mathematical sense even before the invention of the telescope.

An example that illustrates the struggle for a suitable instrument of measurement is Galileo Galilei's investigations of free fall.[4] Even after slowing down the process of falling by relocating the motion to the inclined plane, his time measurements were still inaccurate because he used his own pulse as a standard. Thus, it became necessary to determine the time differences and ratios via the detour of mass differences and ratios, which resulted from the uniform outflow from an elevated vessel.

Today, as well, the concept of quantities as "measurable natural phenomena" (Hischer, 2012, p. 138) is central in physics. Physical quantities in the classroom go far beyond those that are already familiar from everyday use, such as length or time. Research shows that the ability to estimate quantities is directly related to

---

[4] See also Volume 1, Chapter "The Mathematization of Physics Throughout History" (Tran, Nguyen, Krause, & Kraus, 2020).

whether the respective quantity can be perceived with the senses. Thus, not only can basic quantities such as mass, length, and temperature be estimated accurately through intuition, but so can derived quantities such as force and velocity. Complex quantities, on the other hand, for which sensory access is not possible, are predominantly overestimated—often by more than one order of magnitude. (Stinken, 2015; Stinken-Rösner, 2015).

Increasingly, even basic physical quantities are affected by a phenomenon that used to be limited to complex measurements: their determination is carried out with complex measuring instruments whose functions are almost impossible to understand. Digital instruments for voltage or electric current are long-established, as the measuring principle remains hidden for students in most cases. Due to the increasing use of electronic measuring tools, they are also found for basic quantities (e.g., electronic thermometers or calipers[5] with digital displays). These tools obscure the underlying process of comparison with a reference or at least direct the focus towards a purely numerical value (see also the chapter "Numbers, Quantities & Units").

While the use of a "black box variant" is avoidable for some instruments, this option is not available for other measuring instruments—a Geiger counter, for example. Pedagogical benefits can be gained by using a historical approach to reveal that each instrument once represented an "open box" in its early use in science (Pinch, 1985).

Also, the hasty introduction of abstract models, such as the interpretation of temperature as the mean kinetic energy of the particles of a system, can have a detrimental effect on the understanding of the quantity in question. Presenting the quantity in the form of a reconstruction based on history can help students understand both the quantity and the process of measurement itself. The intention is not to give a historically accurate presentation of the development of the concept of temperature, but a step-by-step approach, from a qualitative view, via quantitative experiments and laws to embedding in the network of physical theories (Mantyla, 2007).

2.3.2   Equations

Physics today is often associated with the use of formulas or equations, in addition to conducting experiments. However, the concept of the equation as we know

---

[5]A caliper is an instrument used to accurately measure the dimensions of an object, see: https://en.wikipedia.org/wiki/Calipers.

it today is much younger than the natural sciences; even modern physics, established by Galileo Galilei, could not initially make use of the language of formulas. The modern formula notation itself was first established by Leonhard Euler (1707–1783).

Despite the significant advantages of the new notation, it was not used widely in contemporary publications. For example, Isaac Newton explicitly refrained from using the new analytical method out of concern that it would unnecessarily hinder the dissemination of his discoveries among experts. Although the geometric methods were also difficult to understand, their use was widespread, so they at least had the advantage that people were used to them. With Euler, Lagrange, D'Alembert, and Laplace, the analytical approach we are accustomed to today found its way into physical publications (Kuhn, 2016). The original formulations of Newton's *Principia* seem unfamiliar from today's perspective (Fig. 2.1). The further development of mathematics thus caused the phenomenon of social selection, through which participants were excluded from the discourses of natural philosophy (Gingras, 2001).

Newton's work is also suitable for emphasizing the educational value of formula representation due to its compact presentation and easy applicability. For this purpose, a comparison between the (original) formulation and the representation using modern notation is suitable (Tab. 2.1).

Only from Euler onward are the great works accessible and familiar to today's scientists who do not have extensive specialist knowledge of their historical genesis. At this point, it should also be emphasized that Euler himself wrote secondary school textbooks to present the elementary basics of mathematics in a new way. Among the many innovations that Euler introduced into mathematics were the signs for differences Δ, sums Σ, binomial coefficients $\frac{p}{q}$, and the integration limits for the integral sign. Furthermore, the names of the sides of a triangle ABC with the corresponding minuscules a, b, and c, as well as the common notation for the angular functions sine, cosine, and tangent go back to him (Mattmüller, 2010, p. 178–180).

Euler successfully solved a number of complex physical problems with his formula notation. They range from the disturbance to the orbits of comets by the gravitational influence of the planets, the propagation of sound, and the speed of sound, to mechanics in general and shipbuilding in particular—to name but a few examples (cf. Gautschi, 2008).

This development in the representation of physical relationships points to one of the roles of mathematics in physics: mathematics functions here as the language of physics. In addition, Uhden (2012) points out two more roles of

Quod fi pondus p ponderi $\mathcal{P}$ æquale partim fufpendatur filo Np, partim incumbat plano obliquo $p\ G$: agantur $p\ H$, $N\ H$, prior horizonti, pofterior plano $p\ G$ perpendicularis ; & fi vis ponderis p deorfum tendens, exponatur per lineam $p\ H$, refolvi poteft hæc in vires $p\ N$, $H\ N$. Si filo pN perpendiculare effet planum aliquod $p\ \mathcal{Q}$; fecans planum alterum $p\ G$ in linea ad horizontem parallela ; & pondus p his planis $p\ \mathcal{Q}$, $p\ G$ folummodo incumberet ; urgeret illud hæc plana viribus $p\ N$, $H\ N$, perpendiculariter nimirum planum $p\ \mathcal{Q}$ vi $p\ N$, & planum $p\ G$ vi $H\ N$. Ideoque fi tollatur planum $p\ \mathcal{Q}$, ut pondus tendat filum ; quoniam filum fuftinendo pondus jam vicem

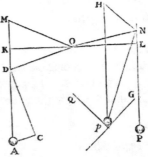

præftat plani fublati, tendetur illud eadem vi $p\ N$, qua planum antea urgebatur. Unde tenfio fili hujus obliqui crit ad tenfionem fili alterius perpendicularis $\mathcal{P}\ N$, ut $p\ N$ ad $p\ H$, Ideoque fi pondus p fit ad pondus A in ratione, quæ componitur ex ratione reciproca minimarum diftantiarum filorum fuorum $p\ N$, $A\ M$ a centro rotæ, & ratione directa $p\ H$ ad $p\ N$; pondera idem valebunt ad rotam movendam, atque ideo fe mutuo fuftinebunt, ut quilibet experiri poteft.

Fig. 2.1   Extract from Newton's Principia. Described and graphically represented is the decomposition of forces. The linguistic representation shown here, based on geometric considerations, is typical of Newton's approach. (Newton, 1726, p. 16, license: public domain)

Tab. 2.1   Overview of different ways of representing the second Newtonian axiom

Representation	Example
Original	Mutationem motus proportionalem esse vi motrici impressae, et fieri secundum lineam rectam qua vis illa imprimitur
Translation	The change of movement is proportional to the applied moving force and follows the direction of the straight line in which that force acts
Formula representation	$\dot{\vec{v}} \propto \vec{F}$

mathematics, through which the different uses and interdependencies of the disciplines can be described. The roles[6] he describes are:

- the pragmatic perspective
- the structural function.

Both functions occur particularly when dealing with equations, especially when considering the educational context. From the pragmatic perspective, mathematics serves as a tool for solving physical problems: often, to students, it is only a matter of finding the appropriate formula to reach a result as quickly as possible.

2.3.3 Functions

The first uses of functions or functional relations can be traced back 4000 years to the combinations of the squares and cubes of the natural numbers. One of their oldest uses is in astronomy, where tables were compiled with positional data that can be understood as a function: position as a function of time. In his astronomical work *Almagest*, Ptolemy also presented the chords of a circle as functions of angles. These tables were calculated and used as an element of the description of nature so that it is obvious to regard them as an empirical source of the function concept. However, the subsequent attribution of the concept of function to these early efforts is an anachronism, since no conceptual mathematical development took place. For Ptolemy, chords were simply lines in a circle, and the transition from angle to chord was not a mathematical process (Kjeldsen, 2015). In addition to tabulated values, early precursors of today's understanding of functions can also be found in the form of function graphs. The earliest representation is from the year 950 (Hischer, 2012, p. 131) and shows the celestial latitude of bodies like the Sun, the Moon, and the planet Saturn, as a function of time (Fig. 2.2). Contrary to modern representations, the horizontal time axis, which is divided 30 times, is not uniform for all celestial bodies; rather, each body is given its own time axis. This means that no temporal relations between the celestial latitudes of the bodies can be derived from the graph (Funkhouser, 1936).

---

[6] In addition, he mentions the "communicative perspective", which we have already discussed under the keyword of language.

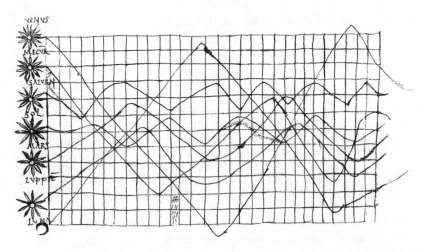

Fig. 2.2 The earliest known graphical representation of a function. It shows the celestial latitude of important celestial bodies as a function of time. (License: public domain)

Nicholas Oresme (1323–1382) also made important contributions to the representation of time-dependent quantities. In his time, Merton College at Oxford was discussing motion sequences and velocities, which in the case of uniformly accelerated motion ultimately led to the so-called Merton rule:

> If a body is uniformly accelerated in time t from the initial velocity v_0 to the final velocity v_1, then the distance travelled is: $s = \frac{v_0 + v_1}{2} t$. (Quoted according to Hischer, 2012, p. 137, Authors' translation)

In contrast to earlier attempts to solve the problem, Oresme used a geometric approach. The visualization of his ideas reveals the functional thinking behind them. By plotting the time on the horizontal axis and the distance travelled by the body on the vertical axis, a Cartesian coordinate system is created here for the first time, and its representation also allows quantitative statements to be made. Historically, Orseme's contribution can be traced back to Galileo Galilei's investigations of free-falling bodies.

However, the real concept of function only came into existence with the invention of the mathematical operation known as differentiation. Thus, the development of the concept of function is directly linked to the development of calculus, which in turn was strongly influenced by physical applications. The term "function" also appears for the first time in a geometrical consideration of curves by

Gottfried Wilhelm Leibniz (1646–1716) (Kjeldsen, 2015). Kjeldsen (2015) summarizes the early development of the term as follows:

> To be sure, the calculus was invented primarily as a means to study curves. After the invention of analytic geometry, which attaches a curve to any equation in two variables, it became necessary to develop means for studying their properties, and so infinitesimal techniques were developed as early as the 1630s. But even in this early period of pre-calculus, curves were often connected to physical phenomena. Indeed, since ancient times some curves (like Archimedes' spiral) had been defined by a continuous motion and after Galileo's investigations of kinematics this became a standard way to think about curves. And as curves were considered from a kinematic point of view, the idea of a variable quantity became a central concept in the emerging field of analysis.

Curves in this context are, as in Euler's sense, lines drawn freehand or mechanically with special instruments, which has no relation to the modern, abstract concept of curves.

The concept of function finally broke away from its close connection to geometry, through the contributions of Johann Bernoulli, among others. Euler gave the following definition in 1748:

> A function of a variable quantity is an analytic expression which is composed in any manner of this variable and of numbers or constants. (quoted from Kjeldsen, 2015)

For Euler, a function was thus a formula to which algebraic operations could be applied, as could transcendental operations such as cosine or sine. Again, prompted by a physical problem—in this case, a vibrating string—Euler, in a discussion with D'Alembert, succeeded in further developing the concept of function (for details see Kjeldsen, 2015).

Of particular interest at this point is the dispute between the two scientists, which can be traced back to their different basic positions. While D'Alembert took the position of an internalist mathematical rigorist, Euler was an externalist for whom mathematics must be designed to solve physical problems.

Again, Kjeldsen and Lützen's (2015) statement on the interaction of mathematics and physics is significant for us:

> So physics forced Euler to extend the concept of function, and it is hard to imagine that such an extension could have been suggested by mathematics itself.

This demonstrates the long tradition of the strongly empirical, problem-solving-oriented use of functions. Hischer (2012, p. 158, Authors' translation) summarizes:

> What is remarkable about all these examples is that (from our point of view) these "functions" were not yet the subject of the respective investigation but were only a "means to a purpose". They were therefore not—as is the case today, for example, in analysis or function theory—the object of consideration, but only supports or even tools for considering a non-mathematical fact.

Time-dependent plots, mainly of movements, were not only the first but are still the most common functional graphs in modern physics education. This is also important for assessing the general educational function of physics, as such time-dependent representations are among the most common forms of data visualization in the media (Hischer, p. 134). Over the course of the book (see the chapter "Functions") it will become clear how closely today's physics teaching is bound to these historical mathematical concepts and practices.

2.3.4 Vectors

The vector has a long tradition as a central mathematical concept. In its history, physical phenomena have played an important role. Thus, it was originally the goal of geometry to describe physical space as accurately as possible (see Struve, 1990). Until the seventeenth century, only scalar quantities were used for this purpose. Over the course of the seventeenth century, physics changed so that quantities such as velocity, force, momentum, and acceleration were used as directed and thus vectorial quantities. Later, the use also spread to electricity, magnetism, and optics (cf. Crowe, 1967).

In this way, the concept of vector developed from the interplay between mathematics and physics. In particular, physical problems played a significant role in terms of motivation:

> The vector calculation was developed in a long historical process, mainly due to the need for a geometric calculation and the requirements of physics. (Filler, 2011, p. 85, Authors' translation)

Historically, one of the most significant physical problems related to vectors was the combination of several forces and velocities acting in different directions. These considerations led to the development of the concept of vectors. At that time, however, the term did not refer to vectors in the modern sense as a class of arrows of equal length and direction—instead, vectors were situated in Euclidean space. The addition of two vectors with the same origin was defined as one vector with the same origin and extending to the opposite corners of the parallelogram defined by the two vectors. Simple ideas about the parallelograms of forces or velocities were employed in ancient Greece and were widely used in the sixteenth and seventeenth centuries, although they were not related to the concept of the vector. Vectors then helped to integrate this approach into a more global concept (cf. Crowe, 1967).

In the further history of the concept of vector in mathematics, ongoing exactifications were made. In particular, the situating of the arrows was omitted and vectors were considered from then on as equivalence classes of parallel arrows of the same length. However, this also had the consequence that certain directed quantities from physics, such as forces, could no longer be simply interpreted as vectors in the mathematical sense (cf. Wittmann, 1996). Nevertheless, further development and formalization up to the concept of vector space led to new fields of application (e.g., in computer science and economics) (cf. Filler, 2011).

The deep historical connections of mathematics and physics in the context of vectors should, according to Laugwitz, lead to integrated teaching:

> The age-honoured motivations for vector addition (parallelogram of forces), of the inner product (work), and of the vector product (moment) should of course be mentioned. They are still useful, if only for euclidean three-dimensional space. (Laugwitz, 1974, p. 245).

2.3.5 Derivations and Integrals

Differential and integral calculus is another mathematical domain that has a strong connection to physics in its historical development. It was developed almost simultaneously and independently by Gottfried Wilhelm Leibniz and Isaac Newton. In this section we want to deal in particular with Newton's theory—a historical excursus of the work of Leibniz is presented in the chapter "Differential Calculus Through Applications" of this volume.

In his work *De methodis serierum et fluxionum*, written in 1670/71, Newton develops an algorithm for quantities that flow over time. This theory, motivated by a physical idea, considers the trajectories of moving objects, such as the motion of a point, which results in a line or the motion of a line, which results in a surface. Newton calls the moving quantities "fluents." Their instantaneous velocities are called "luxions." The infinitely small increase in an infinitely small interval of time is called a "moment." According to Newton, moments behave like fluxions because fluxions would remain constant in an infinitely small interval of time; therefore, moment and fluxion are proportional (cf. Jahnke, 2003).

Thus, Newton's idea has a fundamentally physical character and refers to continuous motion. Furthermore, it has multiple applications in physics. For example, he used integral calculus to deduce Kepler's laws from his own laws of motion (Newton's laws). Leibniz's theory of calculus was also based on references to reality, although with fewer explicit references to classical physical problems.

In the time that followed, many physical theories were built on the basis of calculus. For example, Leonard Euler dealt extensively with physical questions and introduced many forms of notation that are still in use today (cf. Kjeldsen & Lützen, 2015). Other important milestones were the emergence of theoretical mechanics in the eighteenth century (particularly by Maupertuis, Euler, and Lagrange) with the principle of minimal effects (today, Hamilton's principle) and the study of boundary value problems in mathematical physics at the beginning of the nineteenth century (particularly by Green, Gauss, and Dirichlet).

The historical connection is still visible today in mathematics education. Physical applications from the field of kinematics can be found in many textbooks and the principle of continuous motion with instantaneous velocity is often used to introduce differential calculus.

2.3.6   Probability and Statistics

Physical measurements are, related to the corresponding mathematical theory, always affected by errors.[7] To move from experimental data alone to a physical theory is therefore impossible from a practical standpoint. Rather, a theory—mathematically formulated and enabling quantifying statements—must exist in advance in order to be able to classify the results of an experiment. We call this interplay of theory and experiment the "experimental method."[8]

---

[7] See also Volume 1, Chapter "On the Relationship between Mathematics and Physics according to Günther Ludwig" (Geppert et al., 2020).

[8] See also Volume 1, Chapter "The Mathematization of Physics Throughout History" (Tran et al., 2020).

However, even this methodology does not permit any quick conclusions with regard to disproving theories. The aforementioned fact that measured values are never exact makes direct conclusions impossible. The comparison between the predictions of the theory and the results of the experiment can take place only after the results have been analyzed with statistical methods. In contrast, Ernest Rutherford formulated his attitude toward the statistical analysis of measured data as follows: "If your experiment needs statistics, you ought to have done a better experiment."

Especially in astronomy and cosmology, it has always been common to draw conclusions based on uncertain data—as it is necessary today, too, contrary to Rutherford's statement. As in many other fields, history shows that there has been a genuine interaction between astronomy and statistics that has fostered the further development of both disciplines (Coles, 2003).

The starting point of mathematical probability theory is often associated with Jacob Bernoulli, who devoted himself, among other things, to game theory. Thus, he worked out the basics of his theory for large numbers, as occur when throwing dice or randomly drawing different-colored balls from a bag. He was aware that the certainty of statistical statements would increase with the number of observations. However, Bernoulli also found it obvious that his observations on probability theory could not be applied to diseases or the weather, where the reasons for certain developments remained hidden (Stigler, 1986, pp. 63–65).

As early as Galileo, the first approaches to the weighting of measured values can be found, in which values with high uncertainties were given a lower weighting. Later, Daniel Bernoulli and John Michell, two astronomers, investigated whether certain celestial phenomena could be reconciled with random patterns. Bernoulli investigated this concerning the inclination of the orbital planes of the planets, while Michell asked the same question regarding the distribution of the stars on the celestial sphere (Sheynin, 1984; Coles, 2003).

Statistics, and subsequently astronomy, underwent a very significant further development with the application of the least-squares method by Carl Friedrich Gauss. This led to the rediscovery of the dwarf planet Ceres, of which only three individual observations had been previously made (Bruno & Baker, 1999). The connection between the least-squares method and astronomy (or the earth sciences) is described by Stigler (1986, p. 16–17) as follows:

> The development of the method of least squares was closely associated with three of the major scientific problems of the eighteen century: (1) to determine and represent mathematically the motion of the moon; (2) to account for an apparently secular (that is, nonperiodic) inequality that had been observed in the motion of the planets Jupiter and Saturn; and (3) to determine the shape or figure of the earth.

Also, it was Lambert Adolphe Jacques Quételet, a scientist who was also an astronomer, who organized the world's first conference on statistics in 1953 (Coles, 2003). Today's physics and astronomy cannot be imagined without the statistical consideration of measurement errors, which has also found its way into the classroom, as exemplified in the chapter "Lesson Plan on Statistics."

David Hilbert's call for an axiomatization of probability theory was equally closely linked to physics. Its temporal proximity to essential progress in the fields of statistical thermodynamics and mechanics allowed him to call probability theory itself a physical science.[9]

2.3.7 Geometrical Concepts

From the beginning, the mathematical subdiscipline of geometry was closely linked to the natural sciences—and physics in particular. For example, the goal of Euclidean geometry was to describe the construction of figures on a drawing sheet as part of reality. For a long time, Euclid's understanding of mathematics was the leading paradigm and both mathematical and scientific theories, such as Newton's mechanics, were constructed *"more geometrico"* (i.e., axiomatically according to Euclid's elements). This understanding of mathematics was extended in the sixteenth century with the development of projective geometry. Projective geometry had the goal of representing three-dimensional objects in perspective. Thus, the projective geometry of the time was still a theory for the explanation of empirical phenomena (cf. Struve, 1990). For this reason, geometry was often understood as part of the natural sciences, as described in the introduction of Moritz Pasch's *Vorlesungen über die neuere Geometrie*:

> Geometrical concepts form a special group within all concepts, which generally serve for the description of the external world; they refer to shape, measure and relative position of the bodies. Between the geometrical concepts, with the addition of numerical concepts, connections arise that are recognized by observation. With this, the point of view is given, which we intend to hold in the following, according to which we see in geometry a part of natural science. (Pasch, 1976, p. 3, Authors' translation)

---

[9] See the chapter "Stochastics with a Focus on Probability Theory".

With the *Foundations of Geometry* by David Hilbert, formalism was introduced into mathematics and the connection to reality was severed (cf. Freudenthal, 1961). The historical connection is nevertheless significant. For example, Hempel distinguishes between pure geometry and physical geometry. He describes physical geometry as follows:

> Historically speaking, at least, euclidean geometry has its origin in the generalization and systematisation of certain empirical discoveries which were made in connection with the measurement of areas and volumes, the practice of surveying, and the development of astronomy. Thus understood, geometry has factual import; it is an empirical science which might be called, in very general terms, the theory of the structure of physical space, or briefly, physical geometry. (Hempel, 1945, p. 12)

The contemporary geometry taught in schools largely corresponds to Euclidean geometry. The objects of mathematics instruction are empirical objects and students develop an empirical understanding of mathematics similar to that of a natural scientist (cf. Burscheid & Struve, 2009). For this reason, the connection to physics and the description of physical space should be targeted in geometry classes.

2.4   Conclusion

As has been shown in these excerpts, the development of mathematics and physics often ran in parallel. Moreover, for a long time, it was not possible to distinguish between physicists and mathematicians, and it was difficult to assign individual research areas to one subject. The sciences experienced a clear separation because of Hilbert's axiomatization, above all, and the scientists who followed. So today we find a separation that is expressed mainly in terms of the way the two disciplines are related to reality:

> Mathematics operates with abstract, strictly defined objects. The most fundamental of these have been inspired by reality, but are simplified and idealized. [...]
>   Physics, by contrast, deals with the real world of inanimate objects creating theories regarding the world order, its regularity and embedded causality. Where it can, physics tries to be as rigorous as mathematics, but quickly finds that this is often impossible and, in a sense, unnecessary. (Galili, 2018)

In the further course of this book, we want to show how this separation of the two subjects—which did not exist in the past—can be overcome in class without disregarding the characteristics of either discipline. For this purpose, however, it is first necessary to become aware of the different perspectives. Such comparisons will be made using the same concrete subject areas that have been discussed in this chapter, which constitute the essential connections between mathematics and physics in the classroom. The authors are guided by the following core idea:

An implication of such comparison could upgrade the simplistic image: mathematics is not a toolkit or the language of physics, although it might serve as such. Nor does mathematics need to be isolated as a metaphor ignoring reality, although it can be. (Galili, 2018)

References

Bruno, L. C., & Baker, L. W. (1999). *Math & Mathematicians: The history of Math discoveries around the world, 1, A–H*. UXL.

Burscheid, H. J., & Struve, H. (2009). *Mathematikdidaktik in Rekonstruktionen. Ein Beitrag zu ihrer Grundlegung*. Franzbecker.

Coles, P. (2003). Statistical cosmollogy in retrospect. *Astronomy and Geophysics, 44*(3), 3.16–3.20. https://doi.org/10.1046/j.1468-4004.2003.44316.x.

Crowe, M. J. (1967). *A history of vector analysis*. Dover Publications.

Filler, A. (2011). *Elementare Lineare Algebra. Linearisieren und Koordinatisieren*. Springer Spektrum.

Freudenthal, H. (1961). Die Grundlagen der Geometrie um die Wende des 19. Jahrhunderts. *Mathematisch-Physikalische Semesterberichte, 7*, 2–25.

Funkhouser, H. G. (1936). A note on a tenth century graph. *Osiris, 1*, 260–262.

Galili, I. (2018). Physics and Mathematics as interwoven disciplines in science education. *Science & Education, 27*(1–2), 7–37. https://doi.org/10.1007/s11191-018-9958-y.

Gautschi, W. (2008). Leonhard Euler: His Life, the Man, and His Works. *SIAM Review, 50*(1), 3–33. www.jstor.org/stable/20454060.

Geppert, J., Krause, E., Nguyen, P. C., & Tran, N. C. (2020). On the relationship between Mathematics and Physics according to Günther Ludwig. In I. Witzke & O. Schwarz (Series Eds.) & S. F. Kraus & E. Krause (Vol. Eds.). *Comparison of Mathematics and Physics education I: Theoretical foundation for interdisciplinary collaboration* (pp. 137–156). Springer (https://doi.org/10.1007/978-3-658-29880-7_8).

Gingras, Y. (2001). What did Mathematics do to Physics? *History of Science, 39*(4), 383–416. https://doi.org/10.1177/007327530103900401.

Hempel, G. (1945). Geometry and empirical science. *The American Mathematical Monthly, 52*(1), 7–17.

Himbert, M. E. (2009). A brief history of measurement. *The European Physical Journal Special Topics, 172*(1), 25–35. https://doi.org/10.1140/epjst/e2009-01039-1.

Hischer, H. (2012). *Grundlegende Begriffe der Mathematik: Entstehung und Entwicklung*. Vieweg+Teubner Verlag. https://doi.org/10.1007/978-3-8348-8632-3.

Hoffmann, D., & Walker, M. (2006). Zwischen Autonomie und Anpassung die Deutsche Physikalische Gesellschaft im Dritten Reich [Between Autonomy and Adaptation the German Physical Society in the Third Reich]. *Physik-Journal, 2006*, 53.

Jahnke, H. N. (Ed.). (2003). *A history of analysis*. American Mathematical Society.

Kjeldsen, T. H., & Lützen, J. (2015). Interactions between Mathematics and Physics: The history of the concept of function—Teaching with and about nature of Mathematics. *Science & Education, 24*(5–6), 543–559. https://doi.org/10.1007/s11191-015-9746-x.

Kuhn, W. (2016). *Ideengeschichte der Physik [History of Ideas in Physics]*. Springer Berlin Heidelberg (https://doi.org/10.1007/978-3-662-47059-6).

Laugwitz, D. (1974). Motivations and linear algebra. *Educational Studies in Mathematics, 5*(3), 243–254.

Mäntylä, T., & Koponen, I. T. (2007). Understanding the role of measurements in creating physical quantities: A case study of learning to quantify temperature in Physics teacher education. *Science & Education, 16*(3–5), 291–311. https://doi.org/10.1007/s11191-006-9021-2.

Mattmüller, M. (2010). Eine neue Art Mathematik zu schreiben (A new way of writing math). In H. Bredekamp & W. Velminski (Eds.), *Mathesis & Graphé: Leonhard Euler und die Entfaltung der Wissenssysteme* (pp. 177–188). De Gruyter (https://doi.org/10.1524/9783050088235.177).

Metzler, G. (2019). Ein "deutscher Sieg"?: Die Verleihung der Nobelpreise 1919 stand im Spannungsfeld von Politik und Wissenschaft. *Physik Journal, 18*(12), 47–50.

Newton, I. (1726). *Philosophiæ Naturalis Principia Mathematica*. Londini.

Pasch, M. (1976). *Vorlesungen über die neuere Geometrie. Die Grundlegung der Geometrie in historischer Entwicklung*. Springer.

Pinch, T. (1985). Towards an analysis of scientific observation: The externality and evidential significance of observational reports in physics. *Social Studies of Science, 15*(1), 3–36. https://doi.org/10.1177/030631285015001001.

Sheynin, O. B. (1984). On the history of the statistical method in Astronomy. *Archive for History of Exact Sciences, 29*(2), 151–199. https://doi.org/10.1007/BF00348245.

Stigler, S. M. (1986). *The history of statistics: The measurement of uncertainty before 1900*. Belknap Press.

Stinken, L. (2015). Schätzkompetenz von Schülerinnen und Schülern in der Sekundarstufe I. *PhyDid B - Didaktik Der Physik - Beiträge Zur DPG-Frühjahrstagung, 0*(0). http://www.phydid.de/index.php/phydid-b/article/view/600.

Stinken-Rösner, L., & Heusler, S. (2015). Measurement Estimation Skills and Strategies of lower grade Students. In J. Lavonen, K. Juuti, J. Lampiselkä, A. Uitto, & K. Hahl (Chairs), *Science Education Research: Engaging Learners for a Sustainable Future*. Proceedings of ESERA 2015, Helsinki.

Struve, H. (1990). *Grundlagen einer Geometriedidaktik*. Bibliographisches Institut.

Tran, N. C., Nguyen, P. C., Krause, E., & Kraus, S. F. (2020). The Mathematization of Physics Throughout History. In I. Witzke & O. Schwarz (Series Eds.) & S. F. Kraus & E. Krause (Vol. Eds.), MINTUS – Beiträge zur mathematisch-naturwissenschaftlichen Bildung. *Comparison of Mathematics and Physics Education I: Theoretical Foundation for Interdisciplinary Collaboration*. Springer (https://doi.org/10.1007/978-3-658-29880-7_6).

Uhden, O. (2012). Mathematisches Denken im Physikunterricht: Theorieentwicklung und Problemanalyse [Mathematical reasoning in physics education: theory development and problem analysis.]. Technische Universität Dresden, Dissertation. Studien zum Physik- und Chemielernen: Vol. 133. Berlin: Logos.

Wittmann, G. (1996). Eine Unterrichtssequenz zum Vektorbegriff in der Sekundarstufe I. *Mathematica Didactica, 19*(1), 93–116.

Educational Use of Ludwig's Methodology Using the Example of the Lens Equation

3

Jochen Geppert

3.1 The Methodology of Günther Ludwig

In this chapter, a part of geometrical optics is introduced along Ludwig's methodology[1] and made accessible for teaching purposes. The construction of a physical theory according to Ludwig's methodology can be divided into the following steps:

i. Delimitation of the so-called basic area by distinguishing so-called real texts, i.e., physical facts (especially in the form of observations and experiments). These facts then form, possibly by using pre-theories, the basic range. We can consider this range to be open, as the potential further experimental experiences are unlimited.
ii. Marking and reading (presentation) of certain real texts (here, experiments) and linguistic formulation of certain real facts, i.e., real relations.
iii. Choice of an appropriate mathematical theory.
iv. Setting up axioms and sharp principles of representation, i.e., the rules according to which real text pieces and real relations are assigned to mathematical terms and relations:

---

[1] See Volume 1, Chapter 'Ludwig'.

J. Geppert (✉)
Mathematics Education, University of Siegen, Siegen, Germany
e-mail: geppert@mathematik.uni-siegen.de

F. Dilling and S. F. Kraus (eds.), *Comparison of Mathematics and Physics Education II*, MINTUS – Beiträge zur mathematisch-naturwissenschaftlichen Bildung, https://doi.org/10.1007/978-3-658-36415-1_3

a. Selecting image sets: Q_ν
b. Selecting the image relations: R_μ with $R_\mu \subseteq Q_{\nu_1} \times Q_{\nu_2} \times \ldots \times \mathbb{R}$
v. Establishing the blurred imaging principles.

Within Ludwig's program, the basis of a physical theory consists of concrete presentable physical processes and objects. By definition, the basic range of real conditions is open, since, in principle, we can create unlimited new experimental experiences that can be described by the theory that is developed.[2] For this reason, the program mentioned above will be repeated in practice, and the basic area will be grasped progressively more thoroughly. This procedure can be used didactically, because the basic area can be successively expanded through new experiments in the classroom, thus simulating the work of research physicists.

3.2   The Basic Area G of Geometrical Optics

Physical theories are only useful in certain areas—for this reason, we start with the description of the scope of geometric optics with homogeneous materials. In principle, all optical problems can be discovered immediately by solving Maxwell's equations with certain boundary conditions. However, this is impossible in practice and approximations must be made. Such an approximation is the field of geometric optics. It is a strong simplification and describes the propagation of light solely by transversally limited plane waves, so-called light rays.

3.2.1   First Real Texts

In this context, light rays are light bundles, the origin of which represents the first real text (see Sect. 1.3) experimentally investigated in class. The following examples, which can and should be supplemented if enough time is available, are only provided schematically and should first be presented in class as pure observations of nature.

3.2.1.1 The Visualization of Light Beams
Light can only be perceived by people when it is inside the eye. This occurs when light comes straight from the light source or is reflected from bodies in the

---

[2] Cf. also Volume 1, chapters 'Nature of science' and 'Development of knowledge'.

manner: "light source – body – eye." Light bundles that pass the eye can only be recognized if they are directed into the eye. In class, light beams can be made visible with chalk dust or artificial fog.

Educational experiments:

Use an experimental lamp with a condenser whose light exit opening is equipped with a tube of black cardboard about 20 cm long. The lamp emits a broad, almost parallel light beam. Place the lamp in such a way that this light beam falls into a container that is black inside. The experiment is shown schematically in Fig. 3.1.

In a darkened room, the light beam is invisible to the naked eye. If you blow chalk dust or other smoke particles into the room, the light beam becomes visible (the light rays are scattered into the eye by the particles).

If you want to investigate the course of the light beam further, you can create a light trace on a screen (Fig. 3.2).

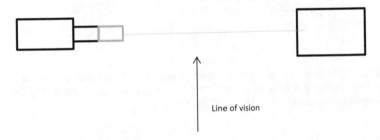

Line of vision

Fig. 3.1 Basic experiment on light perception. A light beam cannot be perceived as long as it does not reach the eye

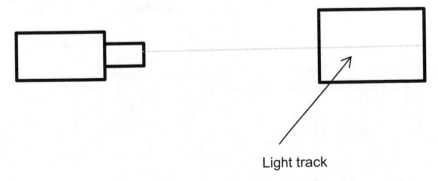

Light track

Fig. 3.2 Basic experiment for generating a light trace on a screen

3.2.1.2 The Rectilinear Propagation of Light

Because no diffraction phenomena are investigated in the context of geometrical optics, it is possible to speak of a straight-line propagation of light even without the restriction of a constant optical density, which is necessary in itself.

Experimentally, the light track above can be traced with a ruler. Shadow images are also useful within this experiment.

The straight-line propagation of the light can be successfully demonstrated with the experiment presented in Fig. 3.3.

The light beam can be made visible by holding a sheet of paper in it so that the light propagates in the plane of the sheet. Light beams can penetrate each other when propagated in a straight line without affecting each other (Fig. 3.4).

Screen

Fig. 3.3 Experiment on straight-line light propagation. Three pinholes (about 3 cm in diameter) are placed in front of an experimental lamp at a distance of about 20 cm in such a way that a narrow beam of light hits a screen vertically

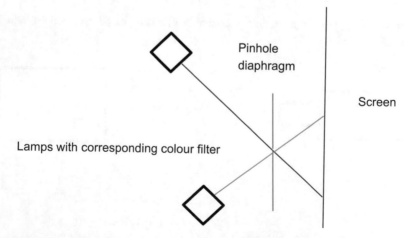

Pinhole
diaphragm

Screen

Lamps with corresponding colour filter

Fig. 3.4 Experiment on the unaffected propagation of two light beams

If a light beam is dropped through a pinhole diaphragm, a light spot is created on a screen in an almost circular illuminated area (which then serves as a light source). It is best to conduct this experiment in a darkened room first and not to make the beam visible from the beginning.

3.2.1.3 Reflection of Light from a Flat Mirror

Depending on the material and type of surface, objects can reflect some of the light falling on them. This process is called reflection. When (almost) all of the light falling on an object is reflected, it is called a specular reflection. Depending on the shape and structure of the reflecting surface, a distinction is made between flat and curved mirrors and between smooth and rough mirrors. Every smooth, flat surface (for example, a smooth metal plate, calm water surface, windowpane, etc.) resembles a flat mirror. While a polished metal surface reflects all of the incident light, glass or water reflects only part of the incident light.

When light is reflected by a plane mirror, the following observations can be made (cf. Fig. 3.5).

Three observations—which do not replace measurements—are that

1. both the incident and reflected light run in a plane perpendicular to the mirror
2. a part of the light is reflected and a part is let straight through the glass
3. the reflected part of the light encloses a right angle with the incident light.

3.2.1.4 Refraction of Light (Snell's law)

A light beam changes its direction when it passes from one medium (such as air) to a second medium of different optical density (glass or water).

The following facts can be demonstrated in class (Fig. 3.6 and 3.7):

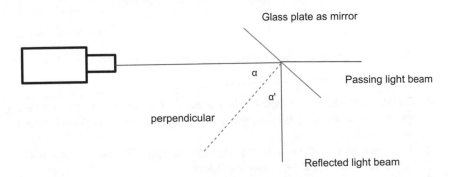

Fig. 3.5 Reflection of a light beam on a flat mirror

refraction to the perpendicular:

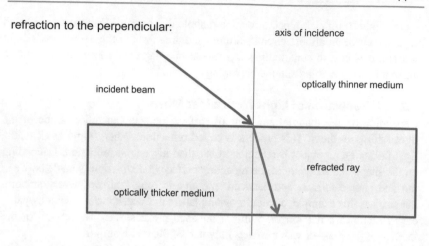

Fig. 3.6 Path of a light beam from an optically thinner to an optically thicker medium

refraction away from the perpendicular:

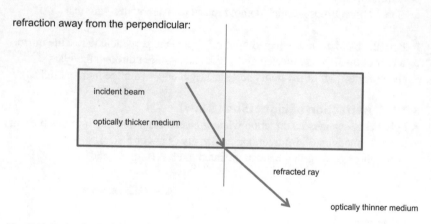

Fig. 3.7 Path of a light beam from an optically thicker to an optically thinner medium

1. The incident and the refracted light bundle lie together in one sheet plane, which is determined by the common solder direction of the transition between the two media.
2. At the transition from the optically denser to the optically thinner medium, the light beam is refracted away from the solder.
3. At the transition from the optically thinner to the optically denser medium, the light beam is refracted towards the solder.

3.2.2 Special Real Texts: Optical Lenses

Cut glass can bundle or diffuse incident light beams depending on the design. Such glass-like bodies, which are usually circular and have a certain curvature, are called "optical lenses." A well-known example is the magnifying glass, which can be used to obtain a magnified image of an object.

Each lens surface can be convex, concave or flat.

- Convex: the surface is curved outwards.
- Concave: the surface is convex towards the inside.
- Flat: the surface is flat.

The course of a beam of light hitting a trapezoidal glass prism can be demonstrated in class.

Here, we can observe a course as it can be recognized schematically in Fig. 3.8.

When the light beam hits the lens section, it is deflected towards the perpendicular on its surface, because the "optical density" of the lens is greater than that

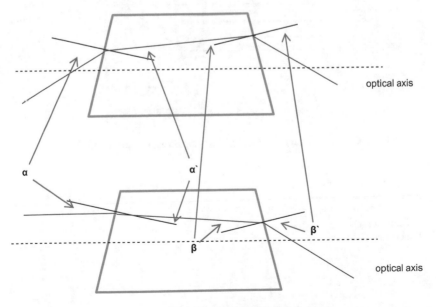

Fig. 3.8 Path of two light beams through a trapezoidal body of glass

of the air surrounding it—the angle α' is smaller than the angle α. Inside the lens, the light beam naturally proceeds in a straight line again until it reaches the opposite side of the lens. There, the beam is refracted a second time. This time, the vertical angle on the back of the lens is greater—β' is greater than β. The overall effect is a significant deflection of the direction of the incoming beam.

While it was previously parallel to the optical axis of the lens, the light beam is now directed towards the optical axis. Consider Fig. 3.9 below as an illustration.

To understand the behavior of the light beam at the converging lens, it is helpful to approximate it with trapezoidal sections, as shown in Fig. 3.10.

To obtain a complete picture, you need to look at all parts of the lens. You will notice that the further "outside" (i.e., away from the optical axis) the rays are deflected, the more they hit the lens.

Collective lenses

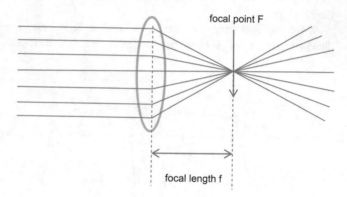

focal point F

focal length f

Fig. 3.9 Schematic illustration of the light focusing through a converging lens

Fig. 3.10 Model of a converging lens. Convex lenses can be regarded as being composed of trapezoidal elements

The rays entering parallel to the optical axis of the converging lens meet at a point, the so-called focal point or focus of the lens. The distance from the lens depends on the strength of the curvature of the two surfaces of the lens and on its optical density, i.e., the material from which it is made. The stronger the curvature of the surfaces and the greater the refractive index of the lens material, the closer the focal point is to the lens and the greater its "refractive power."

3.3   For the Axiomatization of Geometrical Optics

In a first step, only materials with a homogeneous refractive index are considered. This is quite sufficient for school lessons, but it should not be kept secret from the students. The mathematical theory, MT, indicates how the theory can be expanded by showing them new real texts.

3.3.1   The Mathematical Theory—MT

For the description of the image, we choose set theory (Zermelo-Fraenkel axiomatic theory) as a mathematical theory and, based on this, Euclidean-analytical geometry. In this context, the following statement from Euclidean geometry should be emphasized: "The shortest connection between two points A and B lies on the straight line through A and B."

3.3.2   Physical Pre-Theories, Axioms and Sharp Imaging Principles

We posit classical mechanics as a pre-theory for the following considerations.[3]

3.3.2.1 The Path of the Light

From the examples of real texts presented in Chap. 2, we can intuitively form an idea about the propagation of light. We know that light is only visible to the human eye if it travels in such a way that it hits the eye. This is demonstrated by placing a light source in a completely darkened room so that the invisible ray of

---

[3] An introduction to this can be found in Ludwig (1974, Vol. 1).

light falls on a screen in a manner that the bundle of light between the source and the screen is not visible until we push a piece of paper into the previously invisible ray (see Sect. 3.2.1.1).

From this observation, a concept for a "light path" can be determined. As the distances in the presented real texts (experiments) are far too small, we cannot conclude that the speed of light propagation is finite; therefore, we must refer to a corresponding experiment. However, in the case of a finite propagation velocity of the light, nothing speaks against a light path from the above-mentioned "space theory" as a pre-theory.

Here, we have the following mental picture[4] in mind[5]: between a light source and a shade, we situate ourselves at the appropriate places in the room.[6] Small pieces of paper are positioned in front, and we imagine a synchronized clock being attached to each piece of paper. At certain times, the paper discs scatter the light into our eyes, and we can "observe" the propagation of the light in space in this picture.[7] Such a lighting up at a certain time is called a "light event."

In this way, a light path, which the light takes between the source and the screen, can be postulated. However, light events would always be discrete "measured values" (place and time of the respective events), so that the following statement must be made, as the first axiom with regard to mathematics:

Illustration Principle 1

Considering Y in the mathematical picture, we want to define the character set of the light events[8] so that the signs $\vec{A}_1, \ldots, \vec{A}_n$ describe ascertained[9] light events.

---

[4]At this point, it is possible to discuss Einstein's question, which already occupied him as a teenager, of what happens when you run after a beam of light and finally catch it.

[5]To simplify the picture, we can imagine a propagation of light in a vacuum. Light does not need a "carrier," but this is only mathematically described in the wave theory of light.

[6]Which can also be found in a substance.

[7]Of course, it is experimentally impossible to build a camera that takes pictures at intervals and, thus, shows where the light beam is, but it would be conceivable. To illustrate this (the limits of which should be discussed in class), a film can easily be made showing the spread of a water jet from a hose.

[8]Light events naturally depend on the material in which the light is moving—this is taken into account later by the refractive index. The sign Λ represents the set of signs of inertial systems.

[9]It should be emphasized again that we, unfortunately, cannot do this experimentally because of the high speed of light, but in principle, it would be possible.

The mathematical picture continues from $\varrho \subset Y_\lambda$, the set of characters belonging to a light event series, and the (unfortunately only intellectually ascertainable) measurement results in an inertial system (described by the character λ).

A concrete example of this is as follows: at the time $t = 0.004\,s$, light was detected at the space location with the (measured) coordinates $(x = 0.4$ m, $y = 0.6$ m, $z = 1$ m).[10]

The sign of this event (and mathematics can only operate with signs) is then one of the signs A_i.

In the mathematical picture of the Euclidean affine theory, the distance between two points in the number continuum can be determined by the distance formula:

$$\vec{r}_1 := \begin{pmatrix} x_1 \\ y_1 \\ z_1 \end{pmatrix}, \vec{r}_2 := \begin{pmatrix} x_2 \\ y_2 \\ z_2 \end{pmatrix} \Rightarrow d(\vec{r}_1, \vec{r}_2) := \sqrt{(x_1 - x_2)^2 + (y_1 - y_2)^2 + (z_1 - z_2)^2}$$

In the mathematical picture of a physical theory (which we have also assumed above as a pre-theory), we can now set the following sharp mapping principles for two real points marked by the corresponding signs:

$$(-)^1_{lightev.}(1){:}\lambda \in \Lambda, (-)^2_{lightev..}(1) :: \forall i \in \{1, 2, \ldots, n\}, \lambda \in \Lambda{:}\vec{A}_i \in Y_\lambda$$

$$(-)^1_{lightev.}(2){:}R^1_{lightev.}\left(\vec{A}_1, \lambda\right){:}\vec{A}_1 \in Y_\lambda, (-)^2_{lightev.}(2){:}R^2_{lightev.}\left(\vec{A}_2, \lambda\right){:}A_2 \in Y_\lambda,$$

which is to say that the signs $\vec{A}_1 \vec{A}_2$ are two light events in the same inertial system. For example, if you measure a distance of ξm between the two real points in space $\vec{A}_1, \vec{A}_2$, this is written as a relation:

$$(-)^3_{lightev.}(2){:}R^3_{lightev.}\left(\vec{A}_1, A, d\left(\vec{A}_1, \vec{A}_2\right) = \xi\right)$$

The fundamental difference between the experimentally ascertainable reality and this ideal mathematical result is considered in Ludwig's methodology with the so-called "fuzzy mapping principles" (see Sect. 3.3).

---

[10] Ludwig's sets of uncertainty should be omitted at this point to not complicate matters. However, reference should be made to the pre-theories in Ludwig (1974), Vol. 1, Chapter II.

Experimentally, we can see that light propagates with finite speed. It is also important to note in which medium the light moves. Since we will only consider homogeneous materials, which are referred to as mediums in the following, it is sufficient to record the corresponding property of the material as follows:

Axiom 1
"Refractive index of a homogeneous (isotropic) substance".

There is a function that assigns a rational number, called the refractive index (also refered to as the optical density) to each homogeneous substance (which consists of the mass points axiomatically introduced in mechanics and has a constant density function). The refractive index n_M is an optical property of the substance[11] and is defined as the ratio of the speed of light in the present medium c_M (substance) to the speed of light in a vacuum c_0:

$$n_M := \frac{c_M}{c_0}$$

Definition
"Homogeneous Optical Medium"

In this context, a homogeneous optical medium can be defined as a substance in which light can propagate. The speed of propagation usually differs from the speed of vacuum light c. An optical medium is homogeneous if its refractive index is independent of location. It is isotropic if its refractive index is independent of the direction of the propagation of light.

Axiom 2
"Path of light"

In the mathematical picture, $n_M \vec{r}(t)$ describes the path of light between the source and the screen in a homogeneous medium and is a continuous function of the time range postulated as a time interval. We call the image of the (mental) measured values the path of light.

At this point, it should be noted (and this must be discussed in class) that this axiom of consistency of $\vec{r}(t)$ cannot be deduced from experience, since also mentally only a finite number of events $(\vec{r}(t_i), t_i)$ is available.

Ludwig writes about this: "the requirement of consistency of $\vec{r}(t)$ is an idealization for the fact that for 'small' time differences $|t_1 - t_2|$ also small distances $d(\vec{r}(t_1), \vec{r}(t_2))$ are found" (Ludwig, 1974, vol. 1, p. 118, translation by the author).

---

[11] This can also be assigned to atomic structures, especially with crystals.

The requirement for the continuity of the path $n_M \vec{r}(t)$ then means that the quantity ϱ imported above is a continuous curve in Y. $t \to n_M \vec{r}(t)$ is, therefore, a continuous mapping of a time interval in Y to the subset $\varrho \subset Y$, which defines light events.[12]

The further axioms are thus postulated relations between the values $\vec{r}(t)$ and t at different times.

Axiom 3

We demand axiomatically that the light path $\vec{r}(t)$ should be a twice differentiable function.

It should be noted that we do not observe a track of light but rather individual refractions of light at different places in space (if the light is refracted, for example, by dust particles). These refractions appear as a homogeneous beam only due to inertia.

Intuitively, we can now guess from the real texts above that the path of light can be described by a straight line in the mathematical picture, i.e., a linear function.

Axiom 4

By an optical system, we want to understand a real text with at least two space points, whereby one of the space points is to be understood as a light source. Every other space point is then to be understood as a screen. The space comprising all space points should be filled with different optical media.

Illustration Principle 2

Concerning L_W, we want to designate the set of characters of the light paths $\Theta_{\vec{A}_i, \vec{A}_j}$, which the light covers between two light events, in the mathematical image between $\vec{A}_i$ and $\vec{A}_{j}$.

It shall therefore apply:

$$(-)^1_{light\ path}(1){:}\Theta_{\vec{A}_i, \vec{A}_j} = \left\{ \vec{A}_i \in Y_\lambda{:}j = 1, \ldots, k \right\}$$

---

[12] At this point, the width of the light beam must be considered. There are two possibilities, either we carry out the experiments with a LASER and then consider the width of the LASER using Ludwig's unsharp quantities, or we choose a so-called "focal beam" from the light bundle for teaching purposes, which is considered to be representative of the propagation of the beam.

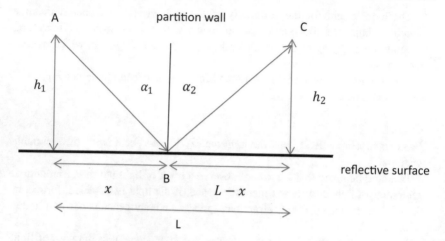

Fig. 3.11 Experiment on light propagation within a homogeneous medium under reflection

$$(-)^2_{lightpath}(1){:}\forall i,j \in \{1,2,\dots,n\}{:}R^1_{lightpath}\left(\vec{A}_i,\vec{A}_j\right){:}\Theta_{\vec{A}_i,\vec{A}_j} \in L_W$$

The length of the light path between $\vec{A}_i$ and $\vec{A}_j$ is then determined by the relation:

$$(-)^1_{lightpath}(2){:}\forall i,j \in \{1,2,\dots,n\}{:}R^2_{lightpath}\left(\vec{A}_i,\vec{A}_j,n_M d\left(\vec{A}_1,\vec{A}_2\right)\right)$$
$$= n_M d\left(A_i,A_j\right)$$

From the observation of the reflection ($\alpha_1 = \alpha_2$) of light, we can intuitively conclude the path of the light beam in the homogeneous medium (Fig. 3.11).

We can intuitively deduce two observations from this: firstly, the form of the path of light, and secondly, the way that the light "chooses" the shortest path. However, this intuitive assumption could not be used to describe the refraction of light, and thus Fermat reformulated it.

Axiom 4

Fermat's principle

In the mathematical picture, the light path in an optically homogeneous medium between two light events is described by a straight line:

$$n_M \vec{r}_L = n_M \left[\begin{pmatrix} A_{1,x} \\ A_{1,y} \\ A_{1,z} \end{pmatrix} + \lambda \cdot \begin{pmatrix} A_{2,x} - A_{1,x} \\ A_{2,y} - A_{1,y} \\ A_{2,z} - A_{1,z} \end{pmatrix}\right], \lambda \in [0,1]$$

Light moves in such a way that it (in a homogeneous matter) "chooses" the straight line that it can pass through in the shortest time.[13]

However, some conclusions can be drawn from this abstract principle and be confirmed experimentally[14]:

1. A plane of incidence of the light exists and is perpendicular to the optical interface. Otherwise, the course of the light—i.e., if the light were to emerge laterally—would contradict Fermat's principle of the shortest time.
2. Light rays that have identical starting and target points have identical optical path lengths.
3. The reversibility of the light path can be deduced from Fermat's principle.

Using this principle of variation, the law of reflection can be easily derived from the mathematical picture.

For the time required for the light beam to travel from A to C via B in a homogeneous medium, we obtain[15]:

$$\Delta t_{ABC} = \frac{\overline{AB}}{c} + \frac{\overline{BC}}{c} = \frac{\sqrt{x^2 + h_1^2}}{c} + \frac{\sqrt{(L - x)^2 + h_2^2}}{c}$$

The condition

$$\frac{dt_{ABC}}{dx} = 0$$

then leads to

$$\frac{dt_{ABC}}{dx} = \frac{x}{c\sqrt{x^2 + h_1^2}} - \frac{L - x}{c\sqrt{(L - x)^2 + h_2^2}} = 0$$

$$cos\left(\frac{\pi}{2} - \alpha_1\right) = cos\left(\frac{\pi}{2} - \alpha_2\right) \Rightarrow \alpha_1 = \alpha_2$$

---

[13] This is not the modern formulation that claims that light moves in such a way that the path is stationary against small changes in the path.

[14] For a more in-depth consideration, see Fouckhardt (1994, p. 22 ff.).

[15] We can suppress the refractive index in the calculation, as it is a constant. The speed of light in a homogeneous medium is also constant, meaning that we can calculate the vacuum speed of light c.

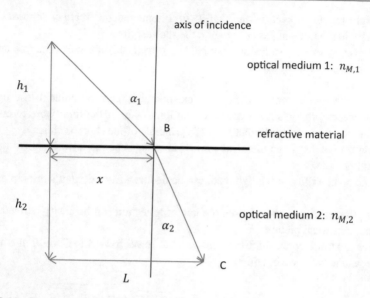

Fig. 3.12 Sketch of the law of reflection according to Snellius

A largely identical calculation prompts the law of refraction according to Snellius (Fig. 3.12).

Here, an analogous calculation leads to

$$\Delta t_{ABC} = \frac{\overline{AB}}{c_{M,1}} + \frac{\overline{BC}}{c_{M,2}} = \frac{\sqrt{x^2 + h_1^2}}{c_{M,1}} + \frac{\sqrt{(L-x)^2 + h_2^2}}{c_{M,2}}$$

with

$$\frac{dt_{ABC}}{dx} = 0$$

and

$$\frac{dt_{ABC}}{dx} = \frac{x}{c_{M,1}\sqrt{x^2 + h_1^2}} - \frac{L-x}{c_{M,2}\sqrt{(L-x)^2 + h_2^2}} = 0$$

and, thus, to the law of refraction according to Snellius:

$$\frac{sin(\alpha_1)}{c_{M,1}} = \frac{sin(\alpha_2)}{c_{M,2}}.$$

3.3.3 Blurred Imaging Principles

3.3.3.1 First Examples: Space Points and Distances

The necessary sharp distinction between the mathematical image and the real texts in Ludwig's methodology means that there are fundamental differences between mathematical and experimental results. Thus, physical space cannot simply be compared with the space $\mathbb{R}^3$ of Euclidean geometry, which as a continuum contains far too many points in space that cannot be physically distinguished under any circumstances.[16]

For this reason, Ludwig provides a structure within the mathematical picture, which, although the experimental possibilities are open in principle, does not deny the indistinguishability of physical results. This will be demonstrated in detail using the example of distance measurement and with the text being based on Ludwig (1974–78). In the mathematical picture of the Euclidean affine theory, the distance between two points in the number continuum can be determined with the distance formula:

$$\vec{r}_1 := \begin{pmatrix} x_1 \\ y_1 \\ z_1 \end{pmatrix}, \vec{r}_2 := \begin{pmatrix} x_2 \\ y_2 \\ z_2 \end{pmatrix} \Rightarrow d(\vec{r}_1, \vec{r}_2) := \sqrt{(x_1 - x_2)^2 + (y_1 - y_2)^2 + (z_1 - z_2)^2}$$

In the mathematical picture of a physical theory (which we have also assumed as a pre-theory), we can now set the following sharp mapping principles for two real points marked by the corresponding signs:

$$(-)^1_{spacepoint}(1){:}\lambda \in \Lambda, (-)^2_{spacepoint}(1){:}\vec{a}_1, \vec{a}_2 \in H_\lambda$$

$$(-)^1_{spacepoint}(2){:}R^1_{spacepoint}(\vec{a}_1, \lambda){:}\vec{a}_1$$
$$\in \lambda, (-)^2_{spacepoint}(2){:}R^2_{spacepoint}(\vec{a}_2, \lambda){:}\vec{a}_2 \in \lambda$$

If we measure a distance of 4.32 m between the two real points in space $\vec{a}_1, \vec{a}_2$, for example, this is written as a relation:

$$(-)^3_{spacepoint}(2){:}R^3_{spacepoint}(\vec{a}_1, \vec{a}_2, d(\vec{a}_1, \vec{a}_2)) = 4.32)$$

However, this relation describes a connection in the mathematical picture that was not found in the real text, since there is a connection in the mathematical

---

[16] Despite the potential for accurate future distance measurement methods, a distance of 10^{-1000} m, for example, will probably remain technically unattainable. However, Ludwig's uncertainty quantities allow such distances in principle if they are experimentally possible.

environment of the two space points (described by $\vec{a}_1, \vec{a}_2 \in H_\lambda$) and further space points (represented again by elements of the "space point character set" H), whose physical prototype cannot be distinguished experimentally from the two real space points. This real-text context must now be considered in the mathematical picture as well. Again, we choose a set of characters $U_H^1 \subset H \times H$ that are connected with the principle of representation. For two real points, with the characters $\vec{a}_i, \vec{a}_j$, that cannot be distinguished experimentally, we write the relation $(\vec{a}_i, \vec{a}_j) \in U_H^1$:

$$(-)^4_{space\,point}(2):(\vec{a}_i, \vec{a}_j) \in U_H^1 \Leftrightarrow R^4_{spacepoint}(\vec{a}_i, \vec{a}_j, B_\mu(\vec{a}_i)):\vec{a}_i, \vec{a}_j \in B_\mu(\vec{a}_i)$$

Another difference between the mathematical picture and the real relation is the mathematical result $d(\vec{a}_1, \vec{a}_2) = 4.32$ and the real result found. If we determine the distance between two real points in space experimentally, we are confronted with the indistinguishability of the points in space, as well as the principally impossible realization of an ideal measurement—between the result of the measurement and the mathematical result $d(\vec{a}_1, \vec{a}_2) = 4.32$, there is a difference, which in turn is considered in an uncertainty quantity and relation.

For the blur set we note:

$$(-)^2_{space\,point}(1):(\vec{a}_1, \vec{a}_2, \alpha), \left(\vec{b}_1, \vec{b}_2, \beta\right) \in U_{H,\mathbb{R}}^2 \subset U_H^1 \times U_d^2, U_d^2$$
$$:= \{(\alpha, \beta) \in \mathbb{R} \times \mathbb{R} : |\alpha - \beta| < \varepsilon\}$$

The corresponding fuzzy mapping principle then consists in the instruction to note, instead of the sharp relation $d(\vec{a}_1, \vec{a}_2)$, the following relation:

$$(-)^5_{space\,point}(2):(\vec{a}_1, \vec{a}_2, \alpha) \in U_{H,\mathbb{R}}^2 : \tilde{R}^1_{space\,point}(\vec{a}_1, \vec{a}_2, \alpha): \Leftrightarrow \exists \vec{b}_1 \in U_H^1, \exists \vec{b}_2$$
$$\in U_H^1, \exists \beta \in \mathbb{R} : d\left(\vec{b}_1, \vec{b}_2\right) = \beta \wedge$$

$$(\vec{a}_1, \vec{a}_2, \alpha), \left(\vec{b}_1, \vec{b}_2, \beta\right) \in U_{H,\mathbb{R}}^2$$

To illustrate this procedure, let us assume that two points in space have been precisely determined and designated by $\vec{a}_1, \vec{a}_2$ and would have determined the distance with $\alpha = 4.32$ m.

According to the blurred illustration principle, a note should now be made:

$$(\vec{a}_1, \vec{a}_2, \alpha) \in U_{H,\mathbb{R}}^2 : \alpha - \varepsilon < d(\vec{a}_1, \vec{a}_2) < \alpha + \varepsilon$$

whereby the two space points are also regarded as smeared, which was considered above by the appropriate relation.

3.3.3.2 Blurred Imaging Principle: Light Events

With Y, we want to designate the character set of the light events in the mathematical picture so that the characters $\vec{A}_1, \ldots, \vec{A}_n$ stand for established light events:

$$(-)^1_{light\ ev.}(1){:}\lambda \in \Lambda, (-)^2_1(1){:}\forall i \in \{1, 2, \ldots, n\}, \lambda \in \Lambda{:}\vec{A}_i \in Y_\lambda$$

As already described above for ordinary points in space, the light events cannot be determined exactly. Therefore, we must write down again:

$$(-)^4_{light\ ev.}(2){:}\left(\vec{A}_i, \vec{A}_j\right) \in \tilde{Y}_\lambda \Leftrightarrow R^3_{light\ ev..}\left(\vec{A}_i, \vec{A}_j, B_\mu\left(\vec{A}_i\right)\right){:}\vec{A}_i, \vec{A}_j \in B_\mu\left(\vec{A}_i\right)$$

For the blur set we note:

$$(-)^2_{light\ ev.}(1){:}\left(\vec{A}_1, \vec{A}_2, \alpha\right), \left(\vec{B}_1, \vec{B}_2, \beta\right) \in Y_{\lambda, \mathbb{R}} \subset \tilde{Y}_\lambda \times U^2_d, U^2_d$$
$$:= \{(\alpha, \beta) \in \mathbb{R} \times \mathbb{R}{:}|\alpha - \beta| < \varepsilon\}$$

Similar to the spatial points belonging to the light events, the associated blurred imaging principle then consists in the statement instead of the exact relation $d(\vec{a}_1, \vec{a}_2)$. This is noted in the following relation:

$$(-)^1_{light\ ev.}(2){:}\left(\vec{A}_1, \vec{A}_2, \alpha\right) \in Y_{\lambda, \mathbb{R}}{:}\dot{R}^1_{light\ ev.}\left(\vec{A}_1, \vec{A}_2, \alpha\right){:} \Leftrightarrow \exists \vec{B}_1$$
$$\in Y_\lambda, \exists \vec{B}_2 \in Y_\lambda, \exists \beta \in \mathbb{R}{:}d\left(\vec{B}_1, \vec{B}_2\right) = \beta \wedge$$

$$\left(\vec{A}_1, \vec{A}_2, \alpha\right), \left(\vec{B}_1, \vec{B}_2, \beta\right) \in U^2_{H, \mathbb{R}}$$

3.3.3.3 Blurred Imaging Principle: Light Path

The path of light consists of a finite number of light events, which we have referred to as a sharp imaging principle:

$$(-)^1_{light\ path}(1){:}\Theta_{\vec{A}_i, \vec{A}_j} = \left\{\vec{A}_i \in Y_\lambda{:}j = 1, \ldots, k\right\}$$

$$(-)^2_{light\ path}(1){:}\forall i, j \in \{1, 2, \ldots, n\}{:}R^1_{light\ path}\left(\vec{A}_i, \vec{A}_j\right){:}\Theta_{\vec{A}_i, \vec{A}_j} \in L_W$$

Since each light event point is now smeared, this results in a smeared light path. We note the following blurred imaging principle:

$$(-)^3_{light\ path}(1) : \left(\vec{A}_i, \vec{A}_j, \alpha\right), \left(\vec{B}_k, \vec{B}_l, \beta\right) \in \tilde{\Theta}_{\vec{A}_i, \vec{A}_j, \mathbb{R}} \subset \tilde{\Theta}_{\vec{A}_i, \vec{A}_j} \times U_d^2, U_d^2$$
$$:= \{(\alpha, \beta) \in \mathbb{R} \times \mathbb{R} : |\alpha - \beta| < \varepsilon\}$$

$$(-)^2_{light\ path}(2) : \left(\vec{A}_i, \vec{A}_j\right) \in \tilde{\Theta}_{\vec{A}_i, \vec{A}_j} \Leftrightarrow R^3_{light\ path}\left(\vec{A}_i, \vec{A}_j, \tilde{\Theta}_{\vec{A}_i, \vec{A}_j}\right) : \vec{A}_i$$
$$\in B_\mu\left(\vec{A}_k\right), \vec{A}_j \in B_\kappa\left(\vec{A}_l\right), \vec{A}_l, \vec{A}_k \in \Theta_{\vec{A}_i, \vec{A}_j}$$

3.3.4 Hypothesis Formation Using the Example of the Lens Equation

3.3.4.1 Images

An optical image is given when rays (often reflected or scattered) emanating from a point of the observed object meet again in a point, which is called an image point due to the influence of the change of the optical medium.

Fermat's principle can also be used to calculate the radiation pattern on thin lenses in the mathematical image.

In this chapter, the hypothesis formation in Ludwig's methodology will be presented through the imaging observation of the converging lens.

For this purpose, a real text (e.g., the image of a candle flame) can be shown in class.

On the screen, we can observe the image of a flame turned upside down (Fig. 3.13).

In the following, the course of the rays in the mathematical picture will be explained using Fermat's principle.

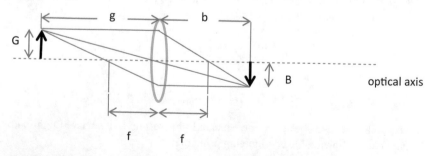

Fig. 3.13 Schematic representation of the inversion of the image with a convex lens. G: Object height, B: Image height, g: object distance, b: image distance and f: focal length

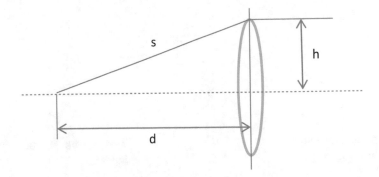

Fig. 3.14 Increase in the length of the light beam path as the distance of the incident light beam from the optical axis increases

3.3.4.2 Derivation of the Lens Equation from Fermat's Principle

To minimize errors in imaging, only light beams close to the axis (so-called paraxial beams) are considered in the following (Fig. 3.14).

We want to investigate the difference in the beam lengths of the oblique beam of length s and the direct beam of length d:

$$d^2 + h^2 = s^2 \Leftrightarrow (s - d)(s + d) = h^2$$

$$s - d = \frac{h^2}{s + d}$$

In the vicinity of the optical axis, the two lengths s and d barely differ, and we can use the approximation $(s + d) \backsim 2s$ to obtain the difference of the lengths:

$$\Delta L \approx \frac{h^2}{2s}$$

According to Fermat's principle, all light paths between source and screen must now have the same time duration.

To simplify the situation, we imagine in the mathematical picture that the converging lens is very thin and only looks at light rays that are close to the axis.

Take another look at the slightly supplemented illustration in Fig. 3.15.

Here, Point D is understood as an image point of point A (which can be seen as a light point source in the mathematical picture).

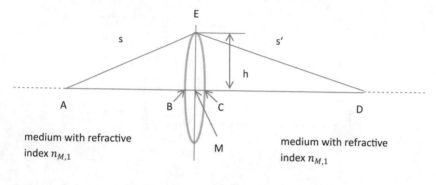

Fig. 3.15 Simplified situation for a thin lens and light beams entering close to the optical axis

According to Fermat's principle, the direct (optical) path ABCD must be as long as the (optical) path over the outermost lens edge M, AED:

$$n_{M,1}\overline{AB} + n_{M,2}\overline{BC} + n_{M,1}\overline{CD} = n_{M,1}\overline{AE} + n_{M,1}\overline{ED} = n_{M,1}\left(\overline{AE} + \overline{ED}\right)$$

Taking into account the above result $s - d \approx \frac{h^2}{2s}$ and $s' - d' = \frac{h^2}{2s'}$ for beams close to the axis ($d = \overline{AM}$ und $d' = \overline{MD}$):

$$\overline{AE} + \overline{ED} \approx \left(\overline{AM} + \frac{h^2}{2\overline{AE}}\right) + \left(\overline{MD} + \frac{h^2}{2\overline{ED}}\right)$$
$$= \overline{AB} + \overline{BM} + \frac{h^2}{2\overline{AE}} + \overline{MC} + \overline{CD} + \frac{h^2}{2\overline{ED}}$$

For the geometrical light path, we must now take into account that BM and MC take place in the glass. The geometrical length difference between the two total paths is:

$$-\Delta := \overline{AE} + \overline{ED} - \left(\overline{AB} + \overline{BC} + \overline{CD}\right)$$
$$\approx \overline{AB} + \overline{BM} + \frac{h^2}{2\overline{AE}} + \overline{MC} + \overline{CD} + \frac{h^2}{2\overline{ED}}$$
$$- \left(\overline{AB} + \overline{BC} + \overline{CD}\right)$$

$$-\Delta = \overline{BM} + \frac{h^2}{2\overline{AE}} + \overline{MC} + \frac{h^2}{2\overline{ED}} - \overline{BC}$$

The resulting difference in optical path length is:

$$-\Delta_L := n_{M,2}\overline{BM} + n_{M,1}\frac{h^2}{2\overline{AE}} + n_{M,2}\overline{MC} + n_{M,1}\frac{h^2}{2\overline{ED}} - n_{M,1}\overline{BC}$$

$$\Delta_L = -\left(n_{M,2} - n_{M,1}\right)\overline{BC} + n_{M,1}\left(\frac{h^2}{2\overline{AE}} + \frac{h^2}{2\overline{ED}}\right)$$

According to the Fermat principle, this difference should now disappear so that

$$\left(n_{M,2} - n_{M,1}\right)\overline{BC} = n_{M,1}\left(\frac{h^2}{2\overline{AE}} + \frac{h^2}{2\overline{ED}}\right)$$

Now, we insert the radius of curvature R of the lens into Fig. 3.16:

As can be seen from the diagram, the following applies:

$$\overline{OM}^2 + h^2 = R^2 \Leftrightarrow h^2 = \left(R + \overline{OM}\right)\left(R - \overline{OM}\right)$$

We can now assume for beams near the axis that

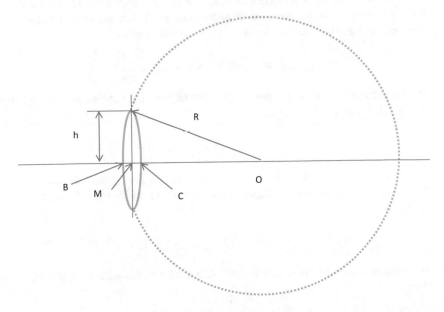

Fig. 3.16 Image supplemented by the curvature of the lens

$$R + \overline{OM} \approx 2R \text{ and } R - \overline{OM} = \frac{1}{2}\overline{BC}$$

and, therefore,

$$h^2 \approx R \cdot \overline{BC} \Leftrightarrow \overline{BC} \approx \frac{h^2}{R}$$

With this result, we obtain, inserted in our above result:

$$(n_{M,2} - n_{M,1})\overline{BC} = n_{M,1}\left(\frac{h^2}{2\overline{AE}} + \frac{h^2}{2\overline{ED}}\right)$$

$$(n_{M,2} - n_{M,1})\frac{h^2}{R} \approx n_{M,1}\left(\frac{h^2}{2\overline{AE}} + \frac{h^2}{2\overline{ED}}\right) \Leftrightarrow (n_{M,2} - n_{M,1})\frac{1}{R}$$

$$\approx n_{M,1}\left(\frac{1}{2\overline{AE}} + \frac{1}{2\overline{ED}}\right)$$

If we now let the distance $\overline{AE}$ increase in thought (this is allowed in the mathematical picture) so that the light beam then (in infinity, so to speak) arrives at the lens axis-parallel, the resulting image point is referred to as the focal point F of the lens. We take the liberty of another approximation, namely:

$$\overline{ED} \approx \overline{DM} = \overline{DC} + \frac{1}{2}\overline{BC} \approx \overline{DC} + \frac{h^2}{2R}$$

We then obtain the following result in the case of parallel light beams close to the axis in the mathematical image:

$$(n_{M,2} - n_{M,1})\frac{1}{R} \approx \frac{n_{M,1}}{2\overline{DM}} =: \frac{n_{M,1}}{2f}$$

whereby the image width $\overline{DM} =: f$. Let us now set $n_{M,1} = 1$, since we start from the medium of air (which we equate with the vacuum by its optical properties), which leads to:

$$(n_{M,2} - 1)\frac{1}{R} \approx \frac{1}{2f}$$

If we use this approximation again, we obtain a relationship called the lens equation:

$$\frac{1}{2\overline{AE}} + \frac{1}{2\overline{ED}} \approx \frac{1}{2f} \Leftrightarrow \frac{1}{\overline{AE}} + \frac{1}{\overline{ED}} \approx \frac{1}{f}$$

3.3.4.3 Derivation of the Imaging Equation from the Lens Equation

Take another look at Fig. 3.17, 3.18.

If we substitute $s = g$ and $s' = b$ into the lens equation, we obtain the imaging equation:

$$\frac{1}{g} + \frac{1}{b} \approx \frac{1}{f}$$

3.3.4.4 Hypothesis Formation using the Example of the Lens Equation I

From the real text of the image of a candle flame, where the heights of the original image and the image on the screen can be estimated using an attached scale, we can postulate that, after drawing the values for G, B, g, b and f on paper and modeling the observed rays with pencil lines, the ray sets also apply to the images

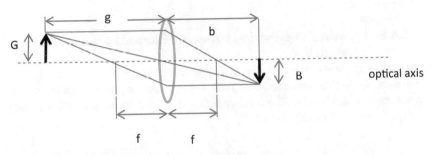

Fig. 3.17 Derivation of the imaging equation from the lens equation. G: Object height, B: Image height, g: object distance, b: image distance and f: focal length

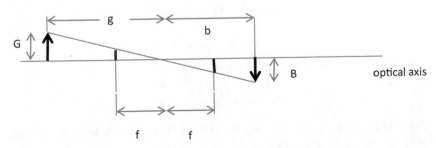

Fig. 3.18 Derivation of the imaging equation from the lens equation in the mathematical picture

of the light rays. This means that in the mathematical image, the following should apply:

$$\frac{B}{G} = \frac{b}{g} \ and \ \frac{B}{G} = \frac{b-f}{f}$$

This leads to:

$$\frac{b}{g} = \frac{b-f}{f} = \frac{b}{f} - 1 \Leftrightarrow \frac{1}{g} = \frac{1}{f} - \frac{1}{b} \Leftrightarrow \frac{1}{g} + \frac{1}{b} = \frac{1}{f}$$

We can now physically interpret this mathematical result, which we obtained from the MT supplemented by the above axioms so that the measured values for the image width b and the object width g can be compared with the focal length of the lens within the error limits in the above relation.

This hypothesis must now be tested experimentally in class. If we find confirmatory results, this affirms the usefulness of our above axioms and, thus, of the entire physical theory.

3.3.4.5 Hypothesis Formation Using the Example of Lens Equation II

An interesting and experimentally verifiable result can be derived from the lens equation. To do this, imagine that $d := b + g = const$, i.e., an attempt is made to place the lens at a fixed distance between the object and the screen so that a sharp image is produced on the screen:

$$\frac{1}{g} = \frac{1}{f} - \frac{1}{b} \Leftrightarrow \frac{1}{b} = \frac{1}{f} - \frac{1}{g} = \frac{g-f}{fg} \Leftrightarrow b = \frac{fg}{g-f} \Leftrightarrow b(g-f) = fg$$

If we now set $d := b + g$ then results by inserting

$$b(d - b - f) = f(d - b)$$

$$b^2 - bd + fd = 0.$$

This way, we get two mathematical solutions:

$$b_{1,2} = +\frac{d}{2} \pm \sqrt{\frac{d^2}{4} - fd}$$

For a solution that can be interpreted in a physically reasonable way, the radicand should be non-negative, so that

$$\frac{d^2}{4} - fd = d\left(\frac{d}{2} - f\right) \geq 0 \Leftrightarrow d \geq 2f$$

In the smaller solution $b_1 = +\frac{d}{2} - \sqrt{\frac{d^2}{4} - fd}$, the image width is smaller than the object width and, therefore, is a reduced image. For the other solution $b_2 = +\frac{d}{2} + \sqrt{\frac{d^2}{4} - fd}$, the image is larger than the object distance and is, thus, an enlarged image.

From the set of equations used in the mathematical image above,

$$\frac{B}{G} = \frac{b}{g} \ and \ \frac{B}{G} = \frac{b-f}{f}$$

results from the specification $d := b + g$ for the two solutions

$$b_1 = +\frac{d}{2} - \sqrt{\frac{d^2}{4} - fd} = g_2 \ \text{and} \ b_2 = +\frac{d}{2} + \sqrt{\frac{d^2}{4} - fd} = g_1$$

and for a 1:1 picture: $g = b = 2f$.

The following experimentally testable hypotheses can, therefore, be formulated from these results obtained in the mathematical picture:

1. For a fixed object-screen distance $d \geq 2f$, there are two settings for the location of the lens: one results in an enlargement of the image and the other in a reduction.
2. In the case of a 1:1 image, the lens must be placed exactly in between the object and the screen.

3.3.4.6 Hypothesis Formation Using the Example of the Lens Equation III

A third verifiable hypothesis can be derived from the lens equation: the rough determination of the focal length of a given lens. Here, it is sufficient to set the object length much larger than the focal length, i.e., $g \gg f$, to obtain the result: $b \approx f$ from the lens equation. Experimentally, we can proceed in such a way that, for example, a ceiling lamp is sharply focused on a screen with a lens, and then the distance between the screen and lens is measured.

3.4   Implementation of Ludwig's Methodology in Teaching

The core goal of physics teaching, in terms of Ludwig's methodology, is to provide answers to the questions, "How does physics arrive at its results? How does physics work?".

3.4.1 Misunderstandings About How Physics Works and How to Avoid Them Within the Framework of Ludwig's Methodology

Ludwig points to various misunderstandings that arise in connection with the way physics works (see in particular Ludwig, 1990, p. 6 ff) and, if not taken into account in class, lead to misconceptions about physics.

Some of these misconceptions are briefly listed below, followed by a discussion of how to avoid them in class using Ludwig's methodology.

3.4.1.1 Physics Does Not Clarify the "why" Question
In this regard, Ludwig states (Ludwig, 1990, p. 6):

> For example, everyone knows that objects fall down to earth; and there are some who hope to learn from a physical theory why this is so in nature. But it is precisely this question that a physical theory does not deal with. Instead, a physical theory, for example, can give us more details about the structure of the processes of falling, the orbits of satellites and the orbits of moons and planets, and we can see how all these processes are subject to one and the same principle of order, which is called the physical theory of Newtonian mechanics and Newtonian law of gravity.
>
> The question of why remains open and is not even asked in the context of a physical theory. When one uses the word "why" in physics, one means something completely different: namely, the question of how a currently existing process is to be classified in the ordering principle of a physical theory.

3.4.1.2 The Purpose of Physics is Not to Make Predictions
There should be a clear distinction between physics lessons and technology lessons, although there is significant overlap. Ludwig points out (Ludwig, 1990, p. 6, translation by the author):

> That technicians use physical theories to build apparatuses that will behave in a certain way in the future is different from the physical theory itself. If, for example, a rocket is sent to the moon or a satellite is sent on a certain orbit, then engineers use Newtonian mechanics here. However, the subject of Newtonian mechanics is not the future course of an orbit, but the orbit as a whole. The question that a physical theory deals with is directed at the structure of the orbit and its classification in a general structural principle. For a physical theory, only the experiments that have actually been carried out are of primary importance, and only secondarily the possible outcome of further experiments.

3.4.1.3 The Purpose of Physics is Not to Perform Causal Analyses

Ludwig claims that (Ludwig, 1990, p. 7, translation by the author):

> A statement like "The cause for the falling of a stone to the ground is the attraction of the earth" is physically empty, because the falling process itself remains one and the same in its quantitative and qualitative structure, no matter whether one describes this process in everyday language with "stone falls down" or "earth attracts stone". Not a single physical theory clarifies a cause-effect relationship, such a relationship is not even made the object of physical analysis; the only question of physics is always directed only at the structure of a process and its classification in the most general structural principles possible.

3.4.1.4 Physics is Not a Method to Derive Logically from Experience

This is probably the most common prejudice, but it also has didactically undesirable consequences if, for example, the discharge law is read off in the form of an exponential function from a few measured values on an electrical capacitor in class. A fundamental aspect is the following error, which Ludwig describes as follows (Ludwig, 1990, p. 7):

> Physics is a system that can be deduced from experience. The fact is, however, that no really significant physical theory has ever been derived from experience. Physical theories are open per se, there is no guarantee that firstly, what is derived from a theory can be found in nature in such a way, and secondly, that contradictions to the derived theory will not even be observed.

3.4.2 Didactic Consequences for Teaching

The aim of any lesson should be to inform students about what is involved in a subject that takes a particular view of the world. Teaching should answer the question of how the subject comes to its specific insights and how work is done there. The natural sciences, with their methodology, are particularly suitable for this purpose. For physics lessons, this means that it is important to observe the fruitful interplay between theory and experiment and to practice this in a way that is comprehensible to the students. We should attempt to give students the opportunity to work like "little physicists" whose goal is to find and describe structures in nature. For this purpose, the metatheory of physics of Günther Ludwig is particularly suitable as a didactic framework.

3.4.2.1 Didactic Commentary: Real Texts

Starting from the "real texts," (i.e., phenomena occurring in nature that can be shown in class) connections are intuitively guessed and then associated with mathematical theories, the provision of which is the task of mathematics teaching. Throughout this process, it is possible and desirable to involve the students in the course of guessing. Thus, the experiments mentioned at the beginning can be used to encourage students to describe commonalities and form analogies ("light ray – water ray") and then discuss in class how the observation can be described so that it can be reported to others who have not (yet) seen the experiments. Within sketches or drawings, sections of the route will then automatically appear, and their meaning and limits must be addressed.

3.4.2.2 Didactic Commentary: Axiomatization

Due to the aforementioned request to describe observations to others who have not seen them, we expect that students will make drawings and sketches to document their experimental observations. The students will likely draw the light rays with a ruler and even consider the straight-line propagation of the light to be virtually characteristic of the light. Thus, it is necessary to discuss the meaning and limitations of this description in the classroom. We can allow these suggestions to be drafted about the similarities but also present the differences between their pencil lines and observations and the ideal lines of mathematical theory. These differences can be used for the conscious introduction of axioms, for example, in the way that a light spot of a certain width is still allowed to be represented as a point in the mathematical picture. On the one hand, the axioms can be introduced by the students on an experimental basis. On the other hand, Fermat's principle and the continuity of the path of light are more likely to be introduced by the teacher.

This process should then lead to the selection of a mathematical theory that is familiar to the students. Again, it can be immediately discussed whether the theory can be clearly identified with the observations or whether there are inevitable inaccuracies or differences between the empirical observations and the mathematical theories they have learned.

3.4.2.3 Didactic Commentary: Fermat's Principle

In physics lessons in middle school, a "Fermat-like" formulation of the description of light can be encouraged by letting the students construct the shortest way from A to C via a place B on the partition wall, if light rays are mapped by lines drawn with a sharp pencil (Fig. 3.19).

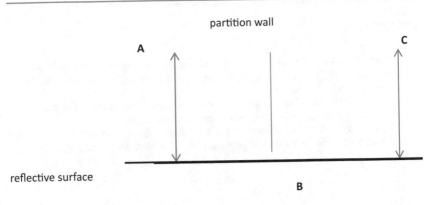

Fig. 3.19   Sketch of the introduction of Fermat's principle for use in school

In upper secondary school lessons, Fermat's principle should then be introduced in the mathematical picture. Here, the advantage of Ludwig's methodology is that confusing formulations such as "light chooses the path of the shortest time" can and should be avoided. Mathematically, while everything is allowed, it must serve to describe and formulate hypotheses in the mathematical picture. Mathematical correlations represent the image of a real process, which in this case is described as if light were to seek different paths and then choose the one with the shortest duration. The limits of this formulation of Fermat's principle should be addressed, as they indicate how openly we should formulate the axioms.

3.4.2.4 Didactic Commentary: Building Hypotheses in the Classroom

The central aim of a series of lessons on geometric optics is, as with any other series of lessons in physics, to develop hypotheses that can be tested. Hypotheses are developed in the mathematical picture of the theory. After having practiced drawing optical experiments on paper (and representing light beams by stretching them), students should find a similarity between a set of rays and the representation of the beam path of a converging lens. When formulating the hypothesis, the students must consider how their mathematical formulation can be observed in reality, what still needs to be considered a confirmation, and what is more likely to be a falsification.

References

Fouckhardt, H. (1994). *Photonik – Eine Einführung in die integrierte Optoelektronik und technische Optik* [Photonics - An introduction to integrated optoelectronics and technical optics]. B.G. Teubner.

Ludwig, G. (1990). *Die Grundstrukturen einer physikalischen Theorie* [The basic structures of a physical theory] (2., überarb. u. erw. Ed.). Springer.

Ludwig, G. (1974–1978). *Einführung in die Grundlagen der theoretischen Physik* [Introduction to the basics of theoretical physics] (4 Vol.). Vieweg Braunschweig.

Equations as a Tool for Hypothesis Formulation in Physics

4

Jochen Geppert

4.1 Summary of the Development of Axiomatic Set Theory

4.1.1 Logic of Statements

Propositional logic is a subfield of logic that deals with propositions and their connection via so-called junctors. The starting points are unstructured elementary statements (atoms), which are assigned a truth value ("true" or "false"). The truth value of a compound statement can then be determined from the truth values of its partial statements without additional information (Schmidt, 1966).

Elements of the propositional logical language
The building blocks of the propositional language are sentence letters ("atomic formulas," sentence constants), punctuators and structuring signs. Phrase letters should be the characters $P_0, P_1, P_2, \ldots$ The punctuation marks should be the characters $\neg, \wedge, \vee, \rightarrow$ and $\leftrightarrow$, which are defined by so-called truth tables (Tab. 4.1).

Round brackets should be used as structuring symbols (A_1). For example, this can be expressed formally in the following way. Let F be the (countably infinite) set of atomic formulas (set letters):

$$F = \{P_n | n \in \mathbb{N}_0\} = \{P_0, P_1 \ldots\}.$$

J. Geppert (✉)
Mathematics Education, University of Siegen, Siegen, Germany
e-mail: geppert@mathematik.uni-siegen.de

© The Author(s), under exclusive license to Springer Fachmedien Wiesbaden GmbH, part of Springer Nature 2022
F. Dilling and S. F. Kraus (eds.), *Comparison of Mathematics and Physics Education II*, MINTUS – Beiträge zur mathematisch-naturwissenschaftlichen Bildung, https://doi.org/10.1007/978-3-658-36415-1_4

Tab. 4.1 Truth table with punctuation marks

A	B	A ∨ B	A ∧ B	A → B	A ↔ B	A	¬A
t	t	t	t	t	t	t	f
t	f	t	f	f	f	f	t
f	t	t	f	t	f		
f	f	f	f	t	t		

Let J_G be the set of punctuators and classification signs: $J_G := \{\neg, \wedge\vee, \rightarrow, \leftrightarrow\}$. The alphabet of logical language is then set to $F \cup J_G$, i.e., the union of atomic formulas, junctors and classification symbols.

4.1.2 Predicate Logic

Predicate logic is an extension of propositional logic. In propositional logic, compound propositions are examined to determine which simpler propositions they are composed of. For example, the statement: "It rains or the Earth is a disk" consists of the two statements: "It rains" and "The Earth is a disk." These two statements cannot be broken down into further partial statements—they are, therefore, called atomic or elementary.

Predicates

A central concept of predicate logic is the predicate. A predicate is a sequence of words with spaces that becomes a true or false statement if a proper name is inserted in each space. For example, the sequence of words "… is a human being" is a predicate because the insertion of a proper name—for example, "Socrates"—results in a statement, such as "Socrates is a human being." With predicate logic, the statement "The Earth is a disk" can be divided into the proper name "the Earth" and the predicate "… is a disk." Based on the definition and the examples, it becomes clear that the term "predicate" in logic, especially in predicate logic, does not have the same meaning as in grammar, even if there is a historical and philosophical connection. Instead of a proper noun, a variable can also be inserted into the predicate, turning the predicate into a sentence function. For example, $\varphi(x) =$ "x is a human being" is a function that, in classical predicate logic, outputs the truth value true for the proper nouns of those individuals who are human beings and the truth value false for all others.

Quantifiers

With quantifiers, statements can be made as to whether a propositional function applies to none, some, or all individuals in the universe of discourse. In the simplest case, the propositional function is a one-digit predicate. If an individual variable is inserted into the predicate, and the existential quantifier and the same variable are placed in front of it, there is at least one individual to which the predicate applies. There must, therefore, be at least one proposition of the form that an individual constant is inserted into the predicate that is true in the universe of discourse concerned. The universal quantifier states that a predicate applies to all individuals in the universe of discourse. Therefore, in classical predicate logic, all atomic, all-quantified statements are true when the universe of discourse is empty. The existential quantifier is expressed in semiformal language as "there is at least one thing so that …" or "there is at least one (variable name) for which …" applies. In formal language, we use the characters $\exists \bigvee$ for the existential quantifier. In semi-formal language, the universal quantifier is expressed as "For all (variable name) applies: …," and in formal language by one of the characters $\forall \bigwedge$.

In formal language, we define the one-numberedness of a predicate by

$$Sing_x(H) := \bigwedge_x \bigwedge_y \left[(H(x) \wedge H(y)) \Rightarrow x = y \right].$$

The use of quantifiers for single-digit predicates is immediately obvious, for example, "_ is a dog." The existentially quantified statement would then be: "There is at least one thing for which the rule applies: it is a dog." In formal language, this would be:

$$\exists x: dog(x) \ or \ \bigvee_x dog(x),$$

where dog(x) is the translation of the single-digit predicate "_ is a dog." Here, the letter x fulfills the function that the word "it" holds in the semi-formal formulation; both denote the space to which the quantifier refers.

To establish the relationship between a quantifier and the space to which it refers, lowercase letters from the end of the Latin alphabet are commonly used, for example, the letters x, y, and z. These are called individual variables. The space to which a quantifier refers, or the variable used to make that connection, is said to be bound by the quantifier. If a blank space in a multi-digit predicate is bound by a quantifier, the result is a predicate one digit lower.

The two-digit predicate $f(\_1, \_2)$ "_1 is father of _2," which expresses the relationship of paternity, becomes a one-digit predicate by binding the first space with the universal quantifier:

$$\forall x: f\left(x, \square_{\square_2}\right) \ or \ \bigwedge_x f\left(x, \square_{\square_2}\right).$$

Speaking to the quality that everyone is a father (which of course is not true), the universal quantifier refers to the first empty space, which is intended for the individual from whom fatherhood originates.

By binding the second blank, however, the one-digit predicate:

$$x: f(\_1, x) \ or \ \bigwedge_x f(\_1, x).$$

This shows that in the property that everyone has a father, the universal quantifier binds the second void, that is, the one intended for the individual who has the role of a son or daughter.

4.1.3 Cantor's Set Theory

In his last paper on set theory, "Contributions to Transfinite Set Theory" (Contributions to the Founding of the Theory of Transfinite Numbers), Mat. Annals 46 (1895) and 49 (1897), Cantor provided the famous definition of a set:

> By a 'set' we understand every summary M of certain, well distinguished objects of our view or our thinking (which are called the 'elements' of M) into a whole.

However, it is impossible to work with this definition, like with Euclid's definition of a point (as something that has no parts). Just as it is not possible to define a point in geometry, we cannot define the term set in set theory either. For example, if we replace the defined "set" in Cantor's definition above with the defining "summary," we see that in no way has something unknown been explained by something known.

Building a mathematical theory on a few, although as minimal as possible, undefined basic concepts is unavoidable. These concepts' relationships to each other and already established mathematical objects are defined.

Therefore, a definition of a set will be omitted in the following, and only some constituent properties will be studied, particularly those that help further the aim of these considerations.

Thus, for each object and set we examine, we should be able to decide whether the element belongs to the set or not. Since we are only dealing with mathematical sets in our context, the first example of a set is a summary of all natural numbers between 1 and 100.

However, we will not describe sets by enumerating their elements but by specifying a property defining the elements of the set, for example:

"x is an element of the set M exactly when the property $H(x)$ applies to the element."

In formal notation:

$$\bigwedge_{x} (x \in M \Leftrightarrow H(x)).$$

Like this:

$$M := \left\{ x \in \mathbb{Z} : (x-5)(x+2)^2 \left(x - \frac{1}{4} \right) = 0 \right\} = \{-2; 5\}.$$

In addition, in a revolutionary statement, Cantor postulated that a set M always exists (i.e., is the object of our view), as long as there is a rule that determines the properties of the elements:

$$\bigvee_{M} \bigwedge_{x} (x \in M \Leftrightarrow H(x)).$$

Herein lies the real novelty of Cantor's definition, since he assumes that potentially infinite sets exist—so-called actual infinite sets. These sets exist without a formation law, such as, for example, the natural numbers created by adding the number 1.

In summary, the essential novelty of the set definition is the arbitrary grouping of objects—this can be referred to as a *naive set formation principle*.

4.1.4 The Russell Typology

With the discovery of antinomies for the Cantorian definition of sets, which simply allows too many freedoms of set formation, such as a set which contains no elements at all:

$$\bigvee_{M} \bigwedge_{x} (x \in M \Leftrightarrow x \notin x) \Rightarrow M := \{\Box\}.$$

This, however, does not lead to contradictions, whereas a similar definition of quantity leads to a blatant contradiction.

The most important antinomy was discovered by B. Russel[1] in 1901 and is briefly described in the following to motivate his typology. Of course, the Cantorian definition of a set does not forbid viewing a set as an element of a set—a set of sets. Consequently, we can easily think of a set as consisting of all subsets of natural numbers from 1 to 100.

Furthermore, a "normal" set does not contain itself as an element. Russel defined normal sets M by the property

$$M \text{ is a normal set} :\Leftrightarrow M \notin M.$$

An example of an "abnormal" set would be the boundless "set of all things." Russel studied such sets and defined the set that is permitted by the Cantorian definition

$$\bigvee_R \bigwedge_M (M \in R \Leftrightarrow H(M): M \notin M),$$

the quantity R of all normal quantities. However, this definition immediately leads to the following:

$$\bigvee_R \bigwedge_M (M \in R \Leftrightarrow H(M): M \notin M) \Rightarrow R \in R \Leftrightarrow R \notin R.$$

This is a logical contradiction, since a statement cannot logically agree with its negation.

This result seems to classify Cantor's set theory as contradictory, but Russell's typology has nevertheless remedied the situation by strictly prohibiting a set definition like the one above. Russell's typology is based on any range of objects, which are called primordial elements and are labeled with Latin letters. State-

[1] Bertrand Arthur William Russell, 3rd Earl Russell (* 18 May 1872 near Trellech, Monmouthshire, Wales; † 2 February 1970 in Penrhyndeudraeth, Gwynedd, Wales) was a British philosopher, mathematician and logician. In 1950, he received the Nobel Prize for Literature. Russell is regarded as one of the fathers of analytical philosophy. He wrote a large number of works on philosophical, mathematical and social topics. Together with Alfred North Whitehead, he published the Principia Mathematica, one of the most important works of the twentieth century on the foundations of mathematics. Source: https://de.wikipedia.org/wiki/Bertrand_Russell—retrieved on 14.02.2020.

ments about these primordial elements are used to form sets in the manner of the Cantorian set theory:

$$M \text{ is a set of first level} :\Leftrightarrow \bigvee_M \bigwedge_x (x \in M \Leftrightarrow H(x)).$$

Quantities are formed, which then, however, are by no means regarded as primordial elements and should be designated, for example, with capital Latin letters. Thus, for example, a "first-level general set" is created by a statement that includes all primal elements. The "first-level sets" formed by the definition above are then again regarded as primal elements from which we again define with statements about these "second-level sets":

$$M \text{ is a set of second level} :\Leftrightarrow \bigvee_{\mathcal{M}} \bigwedge_M \left(M \in \mathcal{M} \Leftrightarrow \widehat{H}(M) \right).$$

This represents so-called quantity systems, the elements of which are first-stage quantities.

We then proceed accordingly and obtain a hierarchy of objects over a given basic range of primordial elements so that every object in this hierarchy has a well-determined step number (including the 0 characterizing the primordial elements).

The general axiom of set formation in Russell's typology is then:

$$y^n \text{ is a set of level } n :\Leftrightarrow \bigvee_{y^n} \bigwedge_{x^{n-1}} (x^{n-1} \in y^n \Leftrightarrow \overline{H}(x^{n-1})).$$

In this case, the upper index designates the number of levels of the type. Statements of the form $x^m \in y^n$ shall be used in a meaningful way only in the case of $m = n - 1$, so that the meaningless statement $x^{n-1} \notin x^{n-1}$ is not possible at any level of the hierarchy $\overline{H}(x^{n-1})$. The set R, on which the antinomy described above is based, can thus no longer be formed, as it would encompass all elements of the hierarchy. However, only so-called "relative subsets" are possible, which combine all objects of the next lower level and whose subsets then belong to elements of the superset of the next higher level.

4.1.5 Zermelo-Fraenkel Set Theory

To bypass Russell's antinomies, Zermelo[2]-Fraenkel[3] set theory only allows delimiting sets of such objects whose range was previously contained in a relative superset $\mathcal{A}$. Only from already given quantities from $\mathcal{A}$ should we, by making meaningful statements $H(x)$, discard new subsets M for elements of these given quantities. These subsets then consist of the elements for which both $x \in \mathcal{A}$ and the statement $H(x)$ apply, so $x \in \mathcal{A} \wedge H(x)$. Thus, Zermelo's initial axiom for quantity formation is

$$\bigwedge_{\mathcal{A}} \bigvee_{M} \bigwedge_{x} \left[x \in M \Leftrightarrow x \in \mathcal{A} \wedge H(x) \right].$$

Comparing this axiom with the Cantorian axiom

$$\bigvee_{M} \bigwedge_{x} (x \in M \Leftrightarrow H(x))$$

we can see that this indicates a limitation of this axiom. Instead of the arbitrary statements in the Cantorian axiom, only those of the type $x \in \mathcal{A} \wedge H(x)$ are approved in Zermelo's axiom.

Thus, the statement $x \notin x$, which occurs in Russell's antinomy, is not possible in this axiom because there is no corresponding superset.[4] This also applies, for example, to the statement permitted in Cantorian set theory: $x \in x \vee x \notin x$. This statement is also not possible here, because again, there is no corresponding superset.

The specification of Zermelo's axiomatic theory, which is now generally accepted as the basis of mathematics, was further developed by Zermelo and Fraenkel and will be presented in the following.

The logical basis of the following axioms is first-level predicate logic with the identity and element predicate $\in$. Zermelo formulated the original axiom system, which is presented below, starting with two definitions for sets and primordial elements:

---

[2] Ernst Friedrich Ferdinand Zermelo (* 27 July 1871 in Berlin; † 21 May 1953 in Freiburg in Breisgau).

[3] Adolf Abraham Halevi Fraenkel, mostly Abraham Fraenkel (* 17 February 1891 in Munich; † 15 October 1965 in Jerusalem).

[4] This would be the set of all sets that do not contain themselves as an element—such an "all set" leads to Russell's antinomy.

$$M \text{ is a set} :\Leftrightarrow (M = \emptyset) \vee \bigvee_x x \in M$$

$$U \text{ is a primordial element} :\Leftrightarrow \neg \bigvee_x x \in U.$$

Thus, sets are objects containing elements, while primordial elements are objects that do not contain elements. According to this definition, the empty set is an excellent primordial element.

1. Extensionality Axiom
Sets are exactly the same if they contain the same elements.

$$\bigwedge_M \bigwedge_{\widetilde{M}} \left(M = \widetilde{M} \Leftrightarrow \bigwedge_x \left(x \in M \Leftrightarrow x \in \widetilde{M} \right) \right)$$

2. Axiom of the Empty Set
There is a set without elements.

$$\exists M_{()}: \bigwedge_x \neg \left(x \subset M_{()} \right)$$

The uniqueness of this set immediately results from the first axiom.

3. Pair Set Axiom
For all objects x and y, there is a set M that contains the elements x and y.

$$\bigwedge_x \bigwedge_y \left[\exists M: \bigwedge_m (m \in M \Leftrightarrow ((m = x) \vee (m = y))) \right]$$

This set is uniquely determined by axiom 1 and is written as $M = \{x, y\}$. The elementary set $M = \{x, x\}$ is also written as $M = \{x\}$.

4. Axiom of Unification
For every set M, there is a set $\widetilde{M}$ that contains exactly the elements of the set M as elements.

$$\bigwedge_M \left[\exists \widetilde{M}: \bigwedge_{\widetilde{m}} \left(\widetilde{m} \in \widetilde{M} \Leftrightarrow \exists m: (m \in M \wedge \widetilde{m} \in m) \right) \right]$$

The set $\widetilde{M}$ is again unambiguous according to axiom 1 and is called a union of the elements of A, written as

$$\widetilde{M} := \cup M.$$

Together with the third axiom, the union of two sets $\overline{M}$ and $\overline{\overline{M}}$ can be defined:

$$\widetilde{M} \cup \overline{\overline{M}} := \cup \left\{\overline{M}, \overline{\overline{M}}\right\}.$$

5. Axiom of Infinity
There is a set M that contains the empty set, and with each element m also the set $m \cup \{m\}$[5]:

$$\exists M: \left[\exists m \in M: \bigwedge_{\widetilde{m}} \neg\left(\widetilde{m} \in m\right) \wedge \left(\bigwedge_m (m \in M \Rightarrow m \cup \{m\} \in M)\right)\right].$$

6. Power Set Axiom
For every set M, there is a set $\mathcal{P}(M)$ whose elements are exactly the subsets of M:

$$\bigwedge_M \left[\exists \mathcal{P}(M): \bigwedge_{\widetilde{M}} \left(\widetilde{M} \in \mathcal{P}(M) \Leftrightarrow \bigwedge_{\widetilde{m}} \left(\widetilde{m} \in \widetilde{M} \Rightarrow \widetilde{m} \in \mathcal{P}(M)\right)\right)\right].$$

This set is clearly determined according to axiom 1.

7. Foundation Axiom[6]
Every non-empty set M contains an element m, so that m and M are disjoint, i.e., m and M have no common element.

$$\bigwedge_M [M \neq \emptyset \Rightarrow \exists m: (m \in M \wedge \neg\exists x: (x \in m \wedge x \in M))]$$

---

[5] There are several such sets, and the intersection of these sets is then the smallest set with these properties and forms the set of natural numbers; the formation of the intersection is generated by axiom 8, and the natural numbers can then be represented by

$$N := \{\emptyset, \{\emptyset\}, \{\emptyset, \{\emptyset\}\}, \{\emptyset, \{\emptyset\}, \{\emptyset, \{\emptyset\}\}\}, \{\emptyset, \{\emptyset\}, \{\emptyset, \{\emptyset\}\}, \{\emptyset, \{\emptyset\}\}\}, \ldots\}.$$

[6] The fundamental axiom prevents a set from containing itself as an element. Such a definition was the reason for Russell's antinomy. It is now not possible that there are infinite or cyclic sequences of sets, each of which contains one in the previous one, for example $M_1 \ni M_2 \ni M_3 \ni \ldots$, because in this case sets like $\widetilde{M} := \{M_1, M_2, M_3, \ldots\}$ are possible, which contradict the axiom: for each $M_i \in \widetilde{M}: M_{i+1} \in M_i \cap \widetilde{M}$.

The element m that is disjoint to M is generally not clearly defined.

8. Axiom of Elimination

A subset exists for each set M, $\widetilde{M} \subseteq M$, which contains exactly those elements m for which the one-digit predicate $P(m)$, in which the character variable $\widetilde{M}$ does not occur, applies:

$$\bigwedge_{M} \left[\exists \widetilde{M} : \bigwedge_{m} m \in \widetilde{M} \Leftrightarrow m \in M \wedge P(m) \right]$$

9. Substitution Axiom

If M is a set and each element of M is uniquely replaced by any set, then M becomes a set.

The replacement by two-digit predicates with properties similar to those of a function can be described somewhat more precisely as an axiom scheme for each predicate:

For each predicate $F(m, n)$ in which the character variable $\widetilde{M}$ does not occur,

$$\bigwedge_{m} \bigwedge_{n} \bigwedge_{p} \left[\begin{array}{c} (F(m,n) \wedge F(m,p) \Rightarrow n = p) \\ \Rightarrow \bigwedge_{M} \left(\exists \widetilde{M} : \bigwedge_{q} \left(q \in \widetilde{M} \Leftrightarrow \exists r : (r \in M \wedge F(r,q)) \right) \right) \end{array} \right]$$

The set $\widetilde{M}$ is clearly determined and is indicated by

$$\widetilde{M} := \left\{ \widetilde{m} \mid m \in M \wedge F\left(m, \widetilde{m}\right) \right\}.$$

If we add the axiom of choice to the above list of Zermelo-Fraenkel's axiomatic, which is abbreviated as "ZF axiomatic," we arrive at "ZFC axiomatic," in which "C" stands for "choice."

10. Axiom of Choice

If M is a set of pairwise disjoint sets, then there is a set that contains exactly one element from each element of M.

Another formulation is:

If M is a set of non-empty sets, then there is a function f of M in its union that assigns an element of M to each element of $\widetilde{M}$ ("selects an element of $\widetilde{M}$"):

$$\bigwedge_{\tilde{m}} \left[\left((\emptyset \in M) \wedge \bigwedge_{\tilde{N},\overline{M},\overline{\overline{M}}} \left[\left(\tilde{N} \in M \wedge \overline{M} \in M \wedge \overline{\overline{M}} \in \tilde{N} \wedge \overline{\overline{M}} \in \overline{M} \right) \Rightarrow \left(\overline{M} = \overline{\overline{M}} \right) \right] \right) \right.$$
$$\left. \Rightarrow \bigvee_{\tilde{M}} \bigwedge_{\tilde{N}} \left(\overline{N} \in M \Rightarrow \exists! \overline{\overline{N}} \colon \left(\overline{N} \in \overline{\overline{N}} \wedge \overline{\overline{N}} \in \tilde{M} \right) \right) \right]$$

The mathematics used in physics can be built on the math used in these axioms of set theory. Ludwig presupposes set theory for every mathematical theory used in physics. In particular, this includes the theoretical possibility of infinite image sets of real objects, although there are only ever a finite number of these in experiments. The point is not to limit the possibilities of physics. Mathematics per se is richer than physics requires. Physics is "finite" in that it only allows a finite number of experiences, but the mathematics it uses must be open. In principle, it must be possible to grasp new discoveries mathematically—for this reason, it is impossible to limit mathematics in physics from the outset, for example, by finite sets, special functions, etc. Furthermore, even a limitation by countable sets is too much of a limitation to the methods of theoretical physics. For example, we can continuously divide any area of space or an object, and only the respective technique limits this state of allocating increasingly finer areas of space. This process should not be restricted mathematically by any axiom of physical theory, but the idealization of infinite quantities, which are now axiomatically sufficiently justified, should be maintained.[7]

4.1.6   Relations and Functions

Under the Cartesian product of two quantities M_1 and M_2, we understand the following set:

$$M_1 \times M_2 := \{(m_1, m_2), m_1 \in M_1, m_2 \in M_2\}$$

Clearly, a relation is a subset of the Cartesian product of two sets:

$$\mathcal{R} \text{ is a relation} :\Leftrightarrow \bigwedge_z \left[z \in \mathcal{R} \Rightarrow \bigvee_x \bigvee_y (z = (x, y)) \right]$$

Then, a function is a unique relation, more precisely a relation that is univalent in the second argument. To keep the function as general as possible, we define a so-called "allclass," which contains all sets (that are allowed) as elements:

---

[7] See chapter "Historical and Educational Relations" in this volume.

$$\mathcal{V} := \{x | Mg(x)\}.$$

The uniqueness of a relation in the second argument is defined by

$$\textbf{\textit{Uniq}}(\mathcal{R}) :\Leftrightarrow \bigwedge_x \bigwedge_y \bigwedge_z \left[(x,y) \in \mathcal{R} \wedge (x,z) \in \mathcal{R} \Rightarrow y = z \right]$$

and, as a function is

$$\textbf{\textit{Fct}}(f) :\Leftrightarrow f \subset \mathcal{V} \times \mathcal{V} \wedge \textbf{\textit{Uniq}}(f).$$

In physics, hypotheses are always expressed in the form of relations and functions, i.e., as mathematical statements about ideal objects, which must then be related to real objects. Depending on the type of hypothesis, these hypotheses are then tested in the form of equations. Therefore, equations play a major role in physics and mathematics.[8]

4.1.7   Equations

Definition: "term".
In mathematics, a term consists of a combination of numbers, variables, symbols for mathematical connections, and brackets, formed according to certain rules. Terms can be seen as the syntactically correct formed words or groups of words in the formal language of mathematics.

In mathematical logic, terms are defined according to the rules by which they are constructed. A term is then any expression that results from applying such rules:

Each variable symbol x is a term.
Each constant symbol k is a term.

If f is an m-digit function symbol and $x_1, \ldots, x_m$ are terms, then $f(x_1, \ldots, x_m)$ is a term as well.

Tarski (1977, p. 21) defines terms as "naming functions," meaning expressions that, after the variables have been replaced by certain constants, become names of things, such as numbers.

---

[8] See Sect. 2.2.1.

For example, the expression $3^x + 5$ is a naming function or term that denotes the number 32 for the assignment $x = 3$.

Schröter (1996, p. 104) defines equations as follows:

Definition: "equation"

1. For two terms S and T, the statement "$S = T$" is called an equation. In particular, a statement of the form $S[x] = T[x]$ is an equation with respect to x.
2. A statement "$S[x] = T[x] \wedge x \in U$" is an equation concerning x in U.
3. A term x represents the solution of the equation $S[x] = T[x]$ in the (mathematical) theory Θ, if $S[x] = T[x]$ can be proved as a proposition within the theory.
4. Accordingly, the term Z represents the solution to the equation $S[x] = T[x] \wedge x \in U$ in the (mathematical) theory Θ, if $S[Z] = T[Z] \wedge Z \in U$ can be proved within the theory as a proposition.

Schröter (1996, p. 104) further remarks:

> According to this definition the concept of equation is very elementary. One does not need any theory to establish an equation, because the equals sign is only an abbreviation. This is opposed by practice. This is because the terms S and T, by which the equation is fixed, are not fixed by unconditional definitions. So you already need a theory to write down the equation at all.

In physics, equations play a central role in the formulation and verification of a hypothesis. In the formulation of a hypothesis, the equation is used as a functional equation, while in the hypothesis, it is used as an equation to be solved.

This method will be applied in the following section on the basis of the hypothesis that an electric charge has a discrete structure, as well as the experimental verification of this hypothesis within the framework of Ludwig's methodology, as described in Ludwig (1990) or Ludwig (1974–1978, vol. 1).

4.2 The Ludwigian Methodology

4.2.1 Pre-Theories of Electrostatics

For the statements of electrostatics described in the following, we need mechanics as physical pre-theories, as described by Ludwig (1974, vol. 1).

4.2.2   Interpretation of the Real Texts of Electrostatics and Their Axiomatization

At this point, we assume that the basic experiments on electrostatics are known. These basic experiments—in Ludwig's language these are the "real texts"—i.e., presentable observations as they are found in nature or artificially produced in the laboratory, outline a first framework of what Ludwig called the "basic range of real conditions." This is considered to be open in principle; this property, which leads to infinite quantities, especially in the mathematical picture, ensures that further experimental experiences are possible.

However, it is important to realize that neither electric charges nor electric fields belong to this basic range. Neither are directly measurable quantities, and they only acquire the status of indirectly measurable quantities through electrostatics.

In summary, the basic experiments on electrostatics are all demonstrations of particular force effects. In Ludwig's methodology, assumptions about the forces can then be made from these real texts. However, a strict derivation of these forces or the following axioms of electrostatics from the real texts is not possible in principle. The real texts of electrostatics show that forces are exerted on small test specimens, which emanate from the media surrounding these bodies.

A close examination of these force effects now reveals the following[9]: the force exerted on a sample at the location $\vec{r}$ by an ambient medium is described by $\vec{k}(\vec{r})$. If we now select other test samples at the same location $\vec{r}$, the forces exerted on these other specimens by the environment show $\vec{k}^j(\vec{r})$, $j = 1, 2, 3, \ldots$, a dependence, which is suggested by the experiments and can be understood as the first axiom of electrostatics[10]:

Axiom (1)

$$\vec{k}^i(\vec{r}) = \lambda(i,j) \cdot \vec{k}^j(\vec{r})$$

The force effect that the environment exerts on a test specimen with the number i at $\vec{r}$ is proportional to the force exerted by the environment on the specimen with

---

[9] For the sake of clarity, we are already using descriptions from the mathematical picture at this point.

[10] In Ludwig's methodology, the axioms are formulated in the mathematical picture—see Ludwig (1990).

the number j. The constant of proportionality does not depend on the location or the environment, but only on the samples. Note that the variables used are images of concrete real texts (in connection with mechanics as a pre-theory), which have this postulated relation. These variables are called signs in Ludwig's methodology in the mathematical picture.

Further experimental experiences then suggest the second axiom:

Axiom (2)

$$\lambda(i,j) = \lambda(i,k) \cdot \lambda(k,j)$$

From this axiom, it follows in the mathematical picture that there must be an unambiguously determined function, except for a constant factor[11]:

$$\lambda(i,j) = \frac{q(i)}{q(j)}.$$

All other solutions $q'(i)$ are obtained from a solution $q(i)$ by means of a constant factor ζ in the form:

$$q' = \zeta q(i).$$

The statement of axiom (1) can then be put into the following form:

$$\vec{k}^i(\vec{r}) = \frac{q(i)}{q(j)} \cdot \vec{k}^j(\vec{r}) \Leftrightarrow \vec{k}^i(\vec{r}) = q(i) \cdot \frac{\vec{k}^j(\vec{r})}{q(j)} := q(i) \cdot \vec{E}(\vec{r}).$$

As will be explained below, Ludwig presents this as a 2nd type hypothesis, because, in the mathematical picture, new variables and signs of an imaginary (presumed) fact are introduced.[12]

The presence of the electric field is postulated even in the absence of the charge, but it can only be measured by its effect on the charge[13] q—whose existence is imaginary. A description of the experimental confirmation of this hypothesis is provided below.

---

[11] This is the same argumentation as Ludwig describes in connection with the measurement of distances and time intervals—see Ludwig (1974–1978, Vol. 1, p. 29 ff.).

[12] In structuralism this axiom corresponds to a Ramsey statement, the theoretical term charge can be determined indirectly from the observation of the orbit (via the second derivative) and the glass rods rubbed with a cat fur (which determine the electric field, for example).

[13] What is actually meant here is a charge function that assigns a charge to a point particle, similar to the mass function (see Ludwig, 1974–1978, Vol. 1), but here, the relationships should not be presented in a too complicated manner.

The following hypothesis can also be derived from the two axioms:

Axiom (1')

$$\vec{k}^i(\vec{r}) = q(i) \cdot \vec{E}(\vec{r})$$

In this case, $\vec{E}(\vec{r})$ is a field that no longer depends on the sample and is uniquely determined, except for one factor.[14]

We call $q(i)$ the electric charge of the ith specimen and $\vec{E}(\vec{r})$ the electric field strength. Experimentally, there are two different types of charges, positive and negative. If the above factor ζ is negative, then all charges and the field strength reverse their signs.

4.2.3   Hypothesis Formation within the Framework of Ludwig's Methodology

In physics, hypotheses are assumptions about the causes of effects or as yet undiscovered objects, as yet unobserved effects, which are generally formulated mathematically. Thus, according to Ludwig, hypotheses are mental speculations in the mathematical picture, which is a reflection of the basic real text in the domain of reality of a physical theory. What is decisive here is the hypothesis, the assumption in a mathematical context. In the mathematical picture, mathematical connections are derived from the axioms of the mathematical theory supplemented by the physical axioms. These connections must then be interpreted in reality—which is not always clear.[15] Most mathematical results are formatted as equations and represent connections between the imaging objects of physical objects.

In his methodology, Ludwig roughly distinguishes between two types of hypotheses.[16] A hypothesis of the first kind consists of the real-text characters, imaginary characters (for which there is no evidence in the real text), and relations between the real-text characters and the imaginary characters. These relations remain within the scope of the real-text evidence considered so far. If this

---

[14] If you now choose q'(i) instead of q(i), you must replace $\vec{E}'(\vec{r}) = \frac{1}{\zeta}\vec{E}(\vec{r})$, so that Axiom (1) in the form $\vec{k}^i(\vec{r}) = q'(i) \cdot \vec{E}'(\vec{r})$ is still preserved.

[15] An example of this would be the wave function in quantum mechanics, the interpretation of which is not clear; in Bohm's mechanics, it is interpreted differently than in the Copenhagen interpretation.

[16] See Ludwig (1974–1978, Vol. 1, p. 87 ff.).

changes, i.e., if new relations between real-text signs and imaginary signs are established, Ludwig calls these "hypotheses of the second kind." The measurement described in the following is the confirmation of a second type of hypothesis of a "granular" or "point-like" charge, as introduced above in Axiom (1'). Axiom (1') is a mathematical derivation of the two axioms (1) and (2). In a mathematical sense, the definition of the two new variables q and $\vec{E}$ is a familiar process that is often introduced for clarification, but in the physical sense, there is more that happens at the same time.

These new variables can only be interpreted in reality as postulates of new physical quantities. here these are the "charge q" and the "electric field $\vec{E}$," whose existence must be verified experimentally. The two new quantities q and $\vec{E}$ are of a different type. The charge q is introduced as a charge function—a function that assigns a charge to a certain mass m independently of its current location, which is understood as discrete and "granular."[17] The properties of an electric charge only become measurable in the presence of an electric field. The electric field, in turn, is a quantity whose existence as an environmental variable can be measured through its force effect on a charge. However, the electric field can be generated independently of a charge used for measurement.

4.2.4 The Determination of e/m in the Millikan Experiment

In this classic experiment,[18] which is shown schematically in the following diagram, very fine oil droplets are blown into the area between the two charged plates via a nebulizer (Fig. 4.1).

The atomization charges the oil droplets electrically. Assuming that a droplet carries the electrical charge of size q and has a mass m, the gravitational force acts in addition to the force of the electric field:

$$\vec{F}_{el} = q \cdot \frac{U}{b} \cdot \vec{e}_z \text{ and } \vec{F}_g = m \cdot g \cdot \vec{e}_z.$$

---

[17] This view is accompanied by interesting and partially still unsolved problems, for example, the reaction of an electron on itself and the related question on how to understand the structure of an electron as a point particle.

[18] The description is based on that given in Brandt and Dahmen (1986, p. 102 ff.). Here, we assume that all physical connections are known.

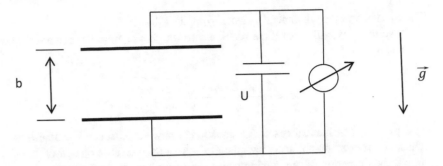

Fig. 4.1 Schematic structure of the Millikan experiment

Depending on how we polarize the voltage, we can set it so that both forces have an opposite direction and their amounts are equal:

$$q \cdot \frac{U}{b} = m \cdot g \Leftrightarrow q = \frac{m \cdot g \cdot b}{U}.$$

The droplets can be considered as small balls with a volume V and a mass m according to:

$$V = \frac{4}{3}\pi \cdot r^3 \Leftrightarrow m = \frac{4}{3}\pi \cdot r^3 \cdot \rho$$

where ρ describes the mass density of the oil. However, the droplets are so small that they cannot be observed directly in the above "hovering phase" under the microscope. This is due to diffraction phenomena.

Instead, we observe the sinking velocity of a single droplet as it sinks when the voltage is switched off and under the influence of gravity and air friction. From mechanics, we know that a mass m falling under friction after a certain time becomes a fall downwards at constant speed[19]:

$$v = \frac{m \cdot g}{R}$$

R represents the coefficient of friction, which is calculated in a viscous liquid like oil according to Stokes' law by

$$R = 6\pi \cdot \eta \cdot r.$$

---

[19] A vivid example of this is the parachute jump.

The constant η is the "thickness" (viscosity) of the oil.

If we then insert the last three relationships into the equilibrium relationship, we obtain

$$q = \frac{6 \cdot \pi \cdot b \cdot \eta \cdot v}{U} \sqrt{\frac{9 \cdot \eta \cdot v}{2 \cdot \rho \cdot g}}.$$

This purely mathematical result (an equation[20]) now represents a hypothesis in which an observation about a new quantity—that of the electric charge of an oil droplet as a function of the sinking speed v between the charged plates—is presented.

We would then have to assume that all variables except v can be regarded as fixed adjustable parameters, i.e., the concrete hypothesis that the measured values can be approximated by a root function.

When experimentally testing this prediction, we see that the function $\rho(v)$ only assumes discrete values and is not similar to the course of the root function. This result can only be interpreted in such a way that there is something like the smallest amount of electric charge.

Brandt and Dahmen write:

It is now astonishing that when measuring many different droplets only charge values occur which are equal to a value of about $1.6 \cdot 10^{-19}$ C or an integer multiple of this value (apart from fluctuations caused by the measurement inaccuracy). (Brandt and Dahmen, 1986, p. 104).

4.3   Practical Consequences for Teaching According to Ludwig's Methodology

The first and most important aim of a cross-curricular approach to teaching mathematics and physics is to reconcile the differences in the contents of the two subjects for the students. The objects of mathematics are ideal and purely mental. Conversely, the objects of physics are real, and knowledge about these objects

---

[20]This equation can be regarded as a function equation. Here, the function value ρ is dependent on the velocity v of the droplet. All other variables can be regarded as parameters that can be set permanently. Surprisingly, the function only takes on discrete values, which we would not expect.

is gained through observations in nature and experiments in the laboratory. The relationship between the objects of mathematics and physics is that of a "reality-picture relationship." A picture shows significant detail, but there are important differences that should not be ignored and should be discussed in class. How exactly a mathematical result corresponds to a real object depends on the technology available, for example, consider the possibilities for producing a steel ball. No matter how technological possibilities develop, we will never reach an ideal state like the mental sphere—the object of mathematics.

For this reason, especially when using geometric models, drawn constructions, or graphic representations in mathematics lessons, their non-ideal character should be emphasized. The central educational value of mathematics lies in the fact that over time, it has detached itself from conceptions and real objects through an axiomatization of its foundations and formalization. This development is partially due to the discovery of contradictions, as described above using the example of Russell's antinomies. Over time, students should carefully gain insight into the power and necessity of mathematics that is deductively based on axioms. In consultation with the physics teacher, mathematical courses can be structured so that this content can be used in physics lessons. Euclidean geometry, for example, should be developed in mathematics lessons to such an extent that the intercept theorems can be used in physics lessons on geometric optics. For a detailed treatment of Newtonian mechanics, infinitesimal calculus should be introduced in mathematics class. Agreements between the teachers of the two subject groups should take this into account.

For physics lessons and the concrete learning of the methods of physics, the use of mathematics means the following: to contend with ideal objects, mathematics must be extended by further axioms, which are supported by observations but cannot be derived from them. At this point, we must point to the "tentative" and "cautious procedure" of this method in the classroom, which is always closely related to an experiment. If we think about nature in terms of physics, we think in ideal objects—mental pictures, which can be described as mathematical objects in a figurative sense. Therefore, if suppositions and hypotheses are to be precise, especially with regard to their limits, they can only be formulated mathematically as equations. A physical hypothesis is motivated by previous experiments and physical axioms. While it can also be formulated freely from the mind, the hypothesis must be tested experimentally.

A fundamental misunderstanding of what constitutes physics would be taking the opposite path and deriving a physical law from the measured values in an inductive way—measured values can motivate a physical connection, but they cannot justify it. If a physical hypothesis is formulated mathematically, we must develop experiments to test it.

References

Brandt, S., & Dahmen, H. (1986). *Physik- Eine Einführung in Experiment und Theorie* (Vol. 2). Springer.

Ludwig, G. (1974–1978). *Einführung in die Grundlagen der theoretischen Physik* (Vol. 4). Vieweg Braunschweig.

Ludwig, G. (1990). *Die Grundstrukturen einer physikalischen Theorie* (2., überarb. u. erw. Ed.). Springer.

Schmidt, J. (1966). *Mengenlehre I*, B.I. Hochschultaschenbücher, Bibliographisches Institut Mannheim.

Schröter, J. (1996). *Zur Meta-Theorie der Physik*. de Gruyter.

Tarski, A. (1977). *Einführung in die Mathematische Logik* (5. Ed.). Vandenhock & Ruprecht.

Comparison: Numbers, Quantities and Units

5

Daniela Götze and Philipp Raack

Dealing with numbers, quantities and units in measurement contexts is a central and also interconnecting content of mathematics and physics classes. Though, many students struggle with measurement tasks. Hence, this chapter analyzes central core insights in measurement-related concepts and procedures focusing on understanding measurement instruments, the importance of reference points, the understanding of unit sizes, converting physical quantities and understanding of decimal numbers. Common problems and mistakes of students are analyzed in detail and some hints how to overcome these problems are given.

5.1 Measurement—A Content for Organizing Reality

One of the central goals of physics classes is to develop conceptual insights in the study of nature and its laws. Such laws help to understand natural phenomena more deeply than a qualitative description because they are based on physical–mathematical equations. Thus, for understanding such laws, students need a profound understanding of physical quantities as an inseparable combination of a number and a specific standardized unit. Gravemeijer et al. (2017) even claim

D. Götze (✉) · P. Raack
Mathematics Education, University of Siegen, Siegen, Germany
e-mail: daniela.goetze@uni-siegen.de

P. Raack
e-mail: raack@physik.uni-siegen.de
Physics Education, University of Siegen, Siegen, Germany

© The Author(s), under exclusive license to Springer Fachmedien Wiesbaden GmbH, part of Springer Nature 2022
F. Dilling and S. F. Kraus (eds.), *Comparison of Mathematics and Physics Education II*, MINTUS – Beiträge zur mathematisch-naturwissenschaftlichen Bildung, https://doi.org/10.1007/978-3-658-36415-1_5

that measurement build bridges between everyday reality and mathematics and dealing with quantities is like quantifying the reality. Hence, measurement as one of the so-called fundamental idea (Vohns, 2002; Winter, 1976) appears continuously in the mathematics and physics curricula in all grades in an age-appropriate manner. Students' abilities concerning educational measurements in upper classes builds upon the education in the lower grades, when children became acquainted with many aspects of measurement (Gravemeijer et al., 2017).

Considering that understanding physical quantities is a core idea of both school subjects, it is astonishing that many students have difficulties to develop a conceptual understanding of quantities and units and that they also struggle with simple measurement tasks, as the following two episodes illustrate.

Episode 1: "We asked fourth graders to write down the body height of their teacher. We found many different estimations: The answers ranged from 40 cm till 4,63 m. Nevertheless, they were able to convert m in km straightforwardly." (Selter & Spiegel, 2010, p. 75)

Episode 2: "Johnny is good at measuring. (…) But one day Johnny's second-grade teacher gave him a ruler that had the 0 marked a short distance from the end of the ruler, and Johnny had trouble. (…) Even after the teacher explained that he must line up the 0 with the endpoint, he was confused. Johnny had learned how to manipulate his ruler and read off the answer without really understanding that his answer represented the distance between the beginning point and the endpoint of the object." (Hiebert, 1984, p. 19)

Both episodes show different but even common problems of students with measuring processes and the interpretation of physical quantities. The first episode shows that many children interpret physical quantities more formally, but they do not have any concrete reference points or conceptual insights for precise estimates. The second episode shows children's problems with typical measuring devices on the one hand, but also with the correct interpretation of a physical quantity like length on the other hand. The question arises how conceptual insights for working flexibly with physical quantities or for combining fundamental quantities to derived quantities can be built on such a rudimental understanding. Besides, these are by far not all of the typical problems of students with measurement, physical quantities and measuring devices. Thus, the aim of this chapter is to give an overview of some central core insights in this mathematical and physical field, on how students struggle if they miss theses core insights and, moreover, to give some hints for overcoming such students' misconceptions.

5.2   Core Insights in Measurement-Related Concepts and Procedures

Obviously, measurement is more than simply determining length, mass, time, temperature (…) or calculating speed, pressure, force (…). Instead of memorizing rules and laws, children up from primary level need a profound understanding of measurement-related concepts and procedures. That means inter alia that they need …

- conceptual insights in using typical *measurement instruments* appropriately and for understanding and validating measuring processes,
- *reference points* for developing a conceptual understanding of physical quantities and their units and for valid measurement estimation,
- a knowledge network for connecting a *broad spectrum of units,*
- insights in *converting physical quantities* and thus in a conceptual *understanding of decimal numbers.*

Obviously, these core insights are often interrelated and connected. Nevertheless, in the following we try to analyze them separately knowing the fact that for solving many mathematical or physical tasks two or more core insights are necessary.

5.2.1   The Complexity of Measuring Instruments

It should be noted that measuring instruments increasingly have "the character of a black box" (Gravemeijer et al., 2017, p. 9). Only a digital timer displays that time is going by, though in a numerically abstract manner. Other digital based instruments like digital scales, digital thermometer or digital distance measuring devices, however, creates as if by magic a physical quantity. The process of (repeated) measuring or measuring in steps using a standardized unit is not apparent. But these measurement experiences are very important for a conceptual understanding of a physical quantity: measuring mass means to find out and to experience the weight of an object, measuring length means to find out and to experience the distance of two points, etc.

Therefore, it is important for teachers to bear in mind that measuring instruments increase the distance to the first-hand experience of nature. Following Wagenschein's beliefs ("save the phenomena!"), measuring devices may be able to detect natural phenomena, but at the same time, they are artificial hurdles for the student who should encounter them for the first time (Wagenschein, 2009). Furthermore, Wagenschein describes the use of modern measuring equipment as "skipped sensuality" (Kircher et al., 2015, p. 53).

The veiling of natural phenomena reaches an abstract maximum with *digital* measuring instruments, since only the *result* is presented. In the course of digitization, more and more non-electrical quantities are measured electrically. Thus, the students get no insights in the underlying physical process, which cannot be in accordance with educational intentions. Hence, it is advisable to discuss the measuring instruments themselves at the beginning of science teaching, including their function and design. The importance of the display type is mostly underestimated in practice, but in an educational setting, it is essential for teachers to know the advantages and disadvantages of analogue and digital displays.

There are two didactically relevant types of measurement displays: scales (e. g. with graduations) and digits-based indicators (numerical displays). Admittedly, a detailed cognitive psychological analysis of both formats cannot be carried out at this point. Nevertheless, the following list is an example of the advantages of analogue measuring instruments in the field of electricity (Volkmer, 1998, p. 30):

1. The scale of an analog multimeter is similar to a length measurement that students become familiar with from an early age.
2. The zero point in the scale center enables conclusions about current direction.
3. The velocity of the indicator movement shows the rate of change.
4. The functioning of the measuring instrument can be explained with magnetism.
5. Fluctuating readings on the digital display confuse the students.

From the last point of the previous list, a different aspect in handling measurements becomes apparent: the unquestioned trust in technology (Kraus, 2017, p. 26). Specifically, this means that students just accept calculator results without validating these result in their minds.

This overreliance on (digital) measurement instrument can also be found in science lessons. Obviously, this seems to be a timeless problem in dealing with technical aids in general: they produce measurement results if by magic and these results must be correct, because digital machines do not make any mistakes. Moreover, knowing the fact that analogue and digital measurement instruments focus different level of abstraction, teachers should be sensitive when and how to use which type of measurement instrument. Figure 5.1 illustrates these different levels of abstraction.

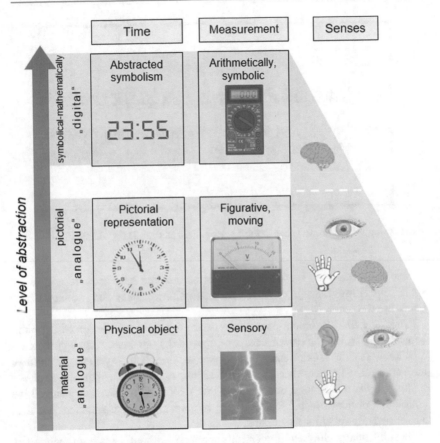

Fig. 5.1 Different levels of abstraction with educational relevance. (Inspired by Leisen, 2010)

Thus, students have to understand how a non-digital measuring instrument works and how to handle it. Those—in the eyes of an experts—trivial things are not trivial for many students. Nührenbörger (2001) requested primary school children to complete a paper strip so that it looks like a ruler. The following documents demonstrate such rulers from third graders who have already learnt length in the second grade. They should have known the essential elements of a ruler (Fig. 5.2).

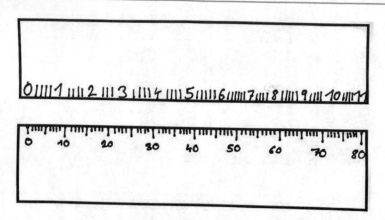

Fig. 5.2 Individual drawn rulers of third graders. (Taken from Dylewski, 2020)

The documents in Fig. 5.2 show that these children have realized and stored some central aspects of a ruler: A ruler has numbers, lines and sometime lines between other lines.

Thus, the first document does not illustrate the close connection of the numbers and the scale. The drawn lines are between the numbers. Although the children know that the "scale" starts with a zero, such rulers indicate that the children have not realized the core idea of length and how to measure length with a ruler.

Rulers, like the second one, indicate that the children may have realized the connection of numbers and scale but such rulers are more like a number line than a ruler because subunits are missing.

Besides, many children note entirely numbers without a scale on their rulers (Nührenbörger, 2001). These children may have not realized the importance and the meaning of a scale. For them a ruler is an object with numbers. Furthermore, many children draw rulers without a zero as a starting point. Such rulers indicate that the children might not have a correct idea of measuring in steps by using a unit and thus starting such measuring processes by zero.

Summarizing, such drawing tasks have a high diagnostical potential, and this idea is conferrable to other measurement instruments like an analog stopwatch, liter measure, analog force measurers, pressure gauge, etc. Such instrument-drawings are very informative for teachers. They help to identify (before or after the content has been treated in mathematics or physics classes) if the students know how special measurement instruments and their scales are constructed, and if they have realized the importance of the zero as starting-point, ... And, generally speaking, correct drawn measurement instruments indicate that the students know how to measure with these instruments (Nührenbörger, 2001).

For fostering these competencies, a meta discourse on how measurement instruments are constructed and, especially, on why they are constructed in this way help (young) students to reach many core insights in measuring processes with measurement instruments (Nührenbörger, 2001).

Nevertheless, this does not mean that they have a realistic idea of the content related units and that they can estimate physical quantities appropriately. For this they need realistic reference points.

5.2.2   The Importance of Reference Points

The episode 1, given at the beginning of this chapter, illustrates that many children have no fundamental idea of units. This means, for example, they do not realize that a teacher cannot have a body height of 40 cm or 4.63 m. They even do not realize that a car cannot weight nearly 1.5 kg (rather 1.5 t), that large objects do not necessarily exert a greater force than small objects, that an edible snail does not crawl with a speed of at least 3 km/h. Such misconceptions are often caused by a lack of understanding of the physical quantity itself and the corresponding unit. Thus, students are not able to estimate length, mass, temperature, speed etc. or they struggle with the correct interpretation, use and connection of the units: A snail crawls with a speed of 3 m/h (not 3 km/h.) But what does this mean? Is a speed of 3 m/h fast? Is it faster than a car or faster than walking?

For such comparative analyzing processes students need realistic reference points (Gravemeijer et al., 2017). Such a reference point may be similar to a standard unit (1 cm as the edge length of most dices, as a children's finger nail, 1 m as the door or panels width, 1 kg as a pack of flour or sugar, ...) or a multiple of a standard unit (2 m as a door height, 1,5 m as span of children's arms, 10 kg as a 10 L bucket full of water, ...). Thus, when making use of a reference point, the physical measurement idea of the reference point object must be compared with the to-be-estimated object. Hence, this to-be-estimated object is frequently not an exact multiple of the reference point object, and final adjustments need to be made (Joram et al., 2005; Nührenbörger, 2001). However, in such estimation processes the students develop a profound understanding about the physical quantity, their units and the iteration of units. Consequently, the main goal of such reference point strategies is to make units more meaningful. Joram et al. (2005) showed in their study among third graders that the use of reference point strategy for length ...

"was statistically associated with greater estimation accuracy (…). In addition, students who estimated using reference points had more accurate representations of standard linear units. (…) When standard units (or multiples of standard units) are represented by reference points, they seem to be more easily recalled and imagined than their corresponding standard units. Having a "feel" for standard linear units, represented by familiar objects, may form a foundation for two-dimensional measurements concepts (area), and three-dimensional measurement concepts (volume)." (Joram et al, 2005, p. 21).

In mathematics as well as in physics classes more attention needs to be given to the development of realistic reference points. For this, reference point objects can, for example, be gathered on different posters: poster for standard units (1 cm, 1 m, 1 km) as well as for multiple of standard units (5 cm, 10 cm, 50 cm). Or they can be collected in small measurement books. Nevertheless, activities of estimating quantities before measuring them exactly can help to deepen the understanding of quantities and units.

5.2.3 Understanding of Units

Measurements of physical quantities are expressed in terms of standardized units. Most common used in the world is the metric based International System of Units (SI). Standardized units help to express, to interpret and to compare measured values in a meaningful way. Though, students frequently misinterpret or ignore units. For them they have an entirely illustrative character for the corresponding numbers (Leuders & Barzel, 2014; Owens & Outhred, 2006). Thus, for fostering a deeper understanding of the core idea of units, students need an intensive confrontation with *semantic* and the *syntactic aspects* of units (Leuders & Barzel, 2014).

Realizing semantic aspects means that the students understand that unit sizes are important for the meaning of the specific number: 3 kg are not the same than 3 g. For understanding this differentiation, reference points become important (see Sect. 1.2.2). They allow to understand that a unit has more than a meaningless illustrating relation to the corresponding number. Moreover, they should know that other unit sizes could be chosen and that the chosen unit size determines the bundling unit of the number. Nevertheless, transforming units normally means to find the most appropriate unit size for the specific physical situation. Therefore, it is important to think and operate with unit sizes flexibly.

The decimal structure of most physical quantities helps to retrieve the bundling unit of specific unit size. Furthermore, parallel word formations (kilo-gramm, kilo-meter) lighten these decimal relationships. As some unit sizes do not follow these analogies (e. g. the bundling of time) such irregularities should be discussed and made transparent for the students.

Realizing syntactic aspects of units means that students are able to operate and calculate with different unit sizes or even combine them. They have to realize that numbers and units change have an inverse relationship: If the bundling unit of the number increases the measurement unit decreases and the other way around. This contrary process causes many irritation (Owens & Outhred, 2006) and will be discussed in the next section.

5.2.4   Converting Physical Quantities and Understanding Decimal Numbers

Converting physical quantities does often imply a conceptual understanding of decimal numbers and, thus, understanding place value. However, arithmetical tasks in primary school predominately work with natural numbers and, thus, primary school children cannot and do not have a conceptual understanding of decimal numbers. But in measurement contexts they have to work and interpret decimal numbers. Nevertheless, it cannot be assumed that students in upper grades have this understanding even though they have learnt decimals numbers in mathematics classes.

> "The recent Third International Mathematics and Science Study showed that internationally about a half of 13 year old students could select the smallest decimal number from a multiple choice list of five decimals." (Steinle & Stacey, 1998, p. 548)

Though, using decimal numbers in a measurement context may be considered as a good opportunity for resolving misconceptions and for fostering conceptual understanding for decimals. However, many measurement tasks allow to interpret decimal numbers in a measurement context as a number consisting of two units and in consequence as two natural numbers divided by a decimal point. For example, 12.45 m can in fact be interpreted as 12 m and 45 cm and 10.382 kg can in fact be interpreted as 10 kg and 382 g. Thus, the decimal point is—in the eyes of (primary school) students—a signal for dividing the physical quantity into a unit and a subunit (Steinle, 2004). Besides, many primary schoolbooks in mathematics support this interpretation. However, focusing on this "decimal point divides" understanding students interpret a decimal measurement number

like 3,48 km as 3 km and 48 m. Furthermore, they are totally confused by interpreting measurement numbers like 3,485 m (commonly they interpret it as 3 m
and 485 cm). With such a "decimal point divides" interpretation there is no need
to partition the reference unit into smaller and smaller amounts. Furthermore, if
children separate the decimal part of a number from the integer part have difficulties in understanding the decimal continuity across the point (Hunter et al., 1994).
Hence, students do not see the extension to further subunits of subunits or the fact
that between any two (decimal) measurement numbers there are infinite many
measurement numbers that can be written (Steinle, 2004).

As a consequence, teachers in mathematics and physics classes should pay
more attention on how students' problems in using and interpreting measurement
numbers are maybe caused by misconceptions of decimals numbers. Hence, it is
important to clarify that a decimal measurement number must be interpreted as
a whole number that indicates the unit and the fractional part of this unit sepa-
rated by a decimal point. In consequence, measurement numbers can be described
differently depending on the chosen unit: 2.3456 kg or 2345.6 g or 2345600 mg.
They all represent the same mass. Nevertheless, they use different measurement
units for bundling.

For overcoming such "the point divides" misconception, a kind of place value
chart that represents different measurement units can help to understand different
bundling processes (see Fig. 5.3) and, thus, to understand different opportunities
for decimal written quantities. A pen can help to indicate the position of the decimal point and, thus, the central measurement bundling unit.

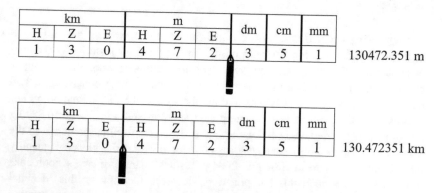

Fig. 5.3 Place value chart for representing central measurement bundling unit

5.3   Central Consequences and Final Remarks

It is helpful to know that there exist some central misconceptions and typical problems of students regarding a conceptual understanding of measurement, physical quantities and units. However, it is important to realize that such errors and mistakes do not signal recalcitrance, ignorance, or the inability to learn (Steinle, 2004). Besides, it is not possible to nip any misconception in the bug. Misconceptions and errors frequently occur. However, a meta discourse about the specific ideas of a physical quantity, about the specific units and the specific measurement instruments can reduce students' misconceptions about numbers, quantities and units and can support meaning making processes.

References

Dylewski, S. (2020). *Zum individuellen Mess- und Längenverständnis von Grundschulkindern.* Unveröffentlichte Bachelorarbeit Universität Siegen.

Gravemeijer, K., Stephan, M., Julie, C., Lin, F.-L., & Ohtani, M. (2017). What mathematics education may prepare students for the society of the future? *International Journal of Science and Mathematics Education, 15*(1), 105–123. https://doi.org/10.1007/s10763-017-9814-6.

Hunter, J., Turner, I., Russell, C., Trew, K., & Curry, C. (1994). Learning multi-unit number concepts and understanding decimal place value. *Educational Psychology, 14*(3), 269–282. https://doi.org/10.1080/0144341940140302.

Hiebert, J. (1984). Why do some children have trouble learning measurement concepts? *Arithmetic Teacher, 31*(7), 19–24.

Joram, E., Gabriele, A. J., Bertheau, M., Gelman, R., & Subrahmanyam, K. (2005). Children's use of the reference point strategy for measurement estimation. *Journal for Research in Mathematics Education, 36*(1), 4–23.

Kircher, E., Girwidz, R., & Häußler, P. (2015). *Physikdidaktik. Theorie und Praxis.* Springer Spektrum.

Kraus, S. (2017). Die persönliche Gleichung in der Astronomie und ihre didaktischen Implikationen. *PhyDid B – Didaktik der Physik – Beiträge zur DPG-Frühjahrstagung, 0.* Retrived from http://www.phydid.de/index.php/phydid-b/article/view/766/912.

Leisen, J. (2010). *Handbuch Sprachförderung im Fach. Sprachsensibler Fachunterricht in der Praxis; Grundlagenwissen, Anregungen und Beispiele für die Unterstützung von sprachschwachen Lernern und Lernern mit Zuwanderungsgeschichte beim Sprechen, Lesen, Schreiben und Üben im Fach.* Varus.

Leuders, T., & Barzel, B. (2014). Größen, Maße und Messen. In H. Linneweber-Lammerskitten (Ed.), *Reihe Lehren lernen. Fachdidaktik Mathematik: Grundbildung und Kompetenzaufbau im Unterricht der Sek. I und II* (pp. 48–68). Seelze, Zug: Klett Kallmeyer, Klett und Balmer.

Nührenbörger, M. (2001). Children's measurement thinking in the context of length. In G. Törner, R. Bruder, A. Peter-Koop, N. Neill, H. Weigand, & B. Wollring (Eds.), *Developments in mathematics education in German-speaking countries. Selected papers from the annual conference on didactics of Mathematics, Ludwigsburg* (pp. 95–106). http://webdoc.sub.gwdg.de/ebook/e/gdm/2001/Nuehrenboerger.pdf.

Owens, K., & Outhred, L. (2006). The complexity of learning geometry and measurement. In A. Gutiérrez & P. Boero (Eds.), *Handbook of research on the psychology of mathematics education: Past, present and future* (pp. 83–115). Sense.

Selter, C., & Spiegel, H. (1997). *Wie Kinder rechnen.* Klett.

Steinle, V. (2004). Detection and remediation of decimal misconceptions. https://www.researchgate.net/publication/237455958_DETECTION_AND_REMEDIATION_OF_DECIMAL_MISCONCEPTIONS.

Steinle, V., & Stacey, K. (1998). The incidence of misconceptions of decimal notation amongst students in Grades 5 to 10. In C. Kanes, M. Goos, & E. Warren (Eds.), *Teaching mathematics in New Times. Proceedings of the 21st annual conference of the mathematics education research group of Australasia* (Vol. 2, pp. 548–555). MERGA.

Vohns, A. (2002). Das Messen als fundamentale Idee im Mathematikunterricht der Sekundarstufe I. In W. Glatz (Ed.), *Siegener Studien, 61,* 157–174. http://wwwu.uni-klu.ac.at/avohns/pdf/beitrag_vohns.pdf.

Volkmer, M. (1998). Ableseübungen an Skalen von Spannungs- und Strommessern. *Naturwissenschaften Im Unterricht/physik, 47*(9), 30–34.

Winter, H. (1976). Die Erschließung der Umwelt im Mathematikunterricht der Grundschule. *Sachunterricht Und Mathematik in Der Primarstufe, 4,* 337–353.

Wagenschein, M. (2009). *Naturphänomene sehen und verstehen. Genetische Lehrgänge, das Wagenschein-Studienbuch.* hep.

Lesson Plan: Measuring Length

6

Nguyen Phuong Chi

Lesson Title: Measuring length
Abstract: The aim of this unit is to work on the unit "meter" and its relation to the unit "centimeter". First, before measuring the objects, students select objects in the classroom whose length can be meaningfully expressed in meters or centimeters. The students also acquire knowledge about examples for different lengths. In order to deepen their knowledge about the use of a ruler, the students, in a further phase, create their own ruler out of cardboard. The individual realization of the essential properties of a ruler serves as a diagnostic tool for further instruction.

Type of school / Grade	Primary School / Grade 2[1]
Prerequisites	none
Number of periods	2

---

[1] This lesson is intended to be taught in the "practice and experiential" part in the Vietnamese mathematics curriculum.

N. P. Chi (✉)
Faculty of Mathematics and Informatics, Hanoi National University of Education, Hanoi, Vietnam
e-mail: chinp@hnue.edu.vn

© The Author(s), under exclusive license to Springer Fachmedien Wiesbaden GmbH, part of Springer Nature 2022
F. Dilling and S. F. Kraus (eds.), *Comparison of Mathematics and Physics Education II*, MINTUS – Beiträge zur mathematisch-naturwissenschaftlichen Bildung, https://doi.org/10.1007/978-3-658-36415-1_6

91

Objectives	Mathematics – Students know the basic length unit "meter" (m) and the relation between "meter" and other units, such as "centimeter" (cm) – Students can measure lengths of specific objects, using popular measurement instruments, such as rulers Physics – Students understand the process for measuring lengths – Students can choose the appropriate unit for measuring lengths in specific situations

6.1   Methodical Commentary

The methodical focus of the lesson series is the student-active development of an idea of the units meter and centimeter as well as the process for measuring. After doing worksheet 1, students can understand the difference between the two unit sizes, meter and centimeter. They can choose the more appropriate unit for measuring a specific object. For example, a book should be measured in centimeters, and a classroom board should be measured in meters.

This activity in phase 4 helps students to fully understand the process for measuring lengths and apply it in measuring the lengths of specific objects. They also have opportunities to practice choosing the appropriate unit for measuring the length of these objects.

In the second to last section of the sequence, students can be creative in creating and decorating their rulers. However, they should know the essential elements of the rulers (numbers, lines, scales) and their relationship. The items that the students have made should express what they understand by measuring lengths and can thus also serve as a diagnostic tool for further instruction.

6.2   Lesson Plan

No./Time	Stage	Learning activities	Interaction form	Materials/ Resources	Methodological comments
# 1 5 min	Beginning the lesson	Say hello to students and tell them the main content of the lesson: they are going to practice measuring lengths of specific objects	Teacher presents	Board, projector	
# 2 15 min	Motivation	Deliver rulers to students and tell them that rulers are used to measure lengths. Ask students to show the difference between meter and centimeter; which distance is longer? Then give students **worksheet 1** and ask them to fill in the appropriate unit for measuring each object in the list	Teacher presents/Teacher asks, students answer/Individual work	Rulers, worksheets	Build understanding of the two units of meter and centimeter; choose appropriate unit for measuring different objects

No./Time	Stage	Learning activities	Interaction form	Materials/ Resources	Methodological comments
# 3 20 min	Introducing new knowledge	Introduce to students the procedure to measure lengths. Model for students how to measure the length of a book (in centimeters) using a ruler. Remind them that the object they measure must begin at the first mark of the ruler and the ruler should be along the edge of the book. Students work in pairs to follow the procedure of measuring the length of the book Then model for students how to measure the length of a desk (in meters). Show students how to mark the end of the ruler with the tip of their pencils in order to measure more than one meter. Students work in pairs to follow the procedure of measuring the length of the desk	Teacher presents/pairs working	Projector, rulers	Through this activity, students can understand the process of measuring length and they can practice measuring lengths according to the process

No./Time	Stage	Learning activities	Interaction form	Materials/ Resources	Methodologi-cal comments
# 4 20 min	Practice	Divide students in groups. Give them **worksheet 2**, ask them to choose three other objects around their class-room to measure and record the results. They have to choose one object that is more appropri-ately measured in meters and another object that is more appropri-ately measured in centimeters. When they finish, invite one or two groups to share their results and explain how they measure their objects	Groups work-ing	Rulers, worksheets	Build deeper understand-ing of the process of measuring length; prac-tice selecting the appro-priate unit of measure for different objects
# 5 25 min	Application	Distribute pieces of cardboard and scissors for students and ask them to create a ruler for measur-ing lengths from these hard paper pieces. Students work in groups to complete the task	Groups work-ing	Cardboard pieces, scis-sors	Encourage creativity through individual design of rul-ers; essential elements of a ruler (numbers, lines, scales) should be known and used
# 6 5 min	Ending the lesson	Summarize the important content of the lesson. Say goodbye to students	Teacher pre-sents	Board, projector	

Worksheet 1

Name of student: ....

Class: ....

Please choose the appropriate unit (cm or m) for measuring the length of each object in the left column.

Objects	Unit (cm or m)
A pencil	
A classroom board	
An eraser	
A desk	
A book	

Worksheet 2

Group number: ....

Class: ....

Find three objects in the classroom to measure their lengths and record the results in the table below. You should choose at least one object that is more appropriately measured in centimeters and another object that is more appropriately measured in meters.

Objects	Lengths

Comparison: Equations in Mathematics and Physics Education

7

Sascha Hohmann and Felicitas Pielsticker

7.1 Introduction

Today, equations are one of the main linking points between mathematics and physics. This term describes a logical proposition concerning the equality between two expressions E_1 and E_2. The basic structure is:

$$E_1 = E_2$$

Equations can be either true (e.g., $1 = 1$) or false (e.g., $1 = 2$). If at least one of the expressions depends on a variable (or unknown),[1] all values of the variables that satisfy the condition of equality are solutions of the equation (for a more detailed explanation, see Chapter "Equations as a tool for Hypothesis formulation in physics").

[1] The definitions of "equation" differ slightly depending on the source and especially the language. In English, any equality (like $1 = 1$) is called an equation, while an equation ("équation," see www.larousse.fr) in France has to have at least one variable (Marcus & Watt, 2012). Here we consider every equality to be an equation.

S. Hohmann
Physics Education, IPN—Leibniz Institute for Science and Mathematics Education, Kiel, Germany
e-mail: hohmann@leibniz-ipn.de

F. Pielsticker (✉)
Mathematics Education, University of Siegen, Siegen, Germany
e-mail: pielsticker@mathematik.uni-siegen.de

© The Author(s), under exclusive license to Springer Fachmedien Wiesbaden GmbH, part of Springer Nature 2022
F. Dilling and S. F. Kraus (eds.), *Comparison of Mathematics and Physics Education II*, MINTUS – Beiträge zur mathematisch-naturwissenschaftlichen Bildung, https://doi.org/10.1007/978-3-658-36415-1_7

The oldest known writings featuring an early form of equations come from ancient Egypt (between 1700 and 1550 BC); equations were used to calculate practical issues. The Greek mathematician Diophantus, who lived in the third century CE, was the first to use variables and is considered one of the founders of algebra. Starting in the fifteenth century, Arabic mathematicians invented notations closer to the algebra we know today (Marcus & Watt, 2012).

At this time, mathematical equations were not yet used in physics. Galileo Galilei used geometry to describe physics (Kim, 2018). In the following centuries, equations became an increasingly important part of physics. While Isaac Newton performed most derivations geometrically (although algebraic and analytical methods were well known to him), Leonard Euler in particular established the use of equations in physics. Today, it is impossible to think about modern physics without using equations.

This strong connection between mathematics and physics is an important part of the teaching of both subjects: If students are to make deductively obtained quantitative predictions in physics lessons, equations are indispensable. Many exercises in mathematics use material from physics; physics textbooks are filled with mathematical equations. In particular, equations are necessary to understand physics at the intermediate level, usually at university. Buschhüter et al. (2016, partly based on the data of Krause & Reiners-Logothetidou, 1978) show a problematic trend in the mathematical abilities of first-year students in physics in Germany. Since 1978, students' ability to work with equations has decreased significantly, by more than 15%. In particular, students find it much more difficult to work with fractions than they did in the past. The percentage of students who provided correct answers decreased significantly for almost all tasks involving fractions, regardless of the mathematical subject area. This shows that technically correct work with equations is imperative in both math and physics and should be encouraged.

But what does "technically correct" mean? Even if equations have been used in mathematics and physics for a long time, there are differences in application between the disciplines. Redish and Kuo (2015) give an example to show the difference between the thinking of mathematicians and physicians, entitled "Corinne's Shibboleth" (originally by Dray & Manogoue, 2002):

One of your colleagues is measuring the temperature of a plate of metal placed above an outlet pipe that emits cool air. The result can be well described in Cartesian coordinates by the function

$$T(x,y) = k\left(x^2 + y^2\right)$$

where k is a constant. If you were asked to give the following function, what would you write?

$$T(r,\theta) = ? \text{(Redish \& Kuo, 2015, p. 7.3)}$$

Physicians interpret T as a temperature and x and y as Cartesian coordinates with the relation $x^2 + y^2 = r^2$ (Pythagoras). Therefore, they give a function in polar coordinates and answer with $T(r,\theta) = kr^2$. For mathematicians, x and y are arbitrary variables that can be replaced by other unknowns: $T(r,\theta) = k(r^2 + \theta^2)$.

While the physicist's solution might appear wrong to the mathematician (there is no extra condition which says $x^2 + y^2 = r^2$), that of the mathematician might appear wrong to the physicist, because adding the area (r^2) to a (squared) angular measurement is meaningless. So, which solution is right depends on the context.

7.2 Mathematical Education Research

The importance of the structural role of mathematics in physics, in general, and the different applications of equations in mathematics and physics, in particular, has often been the focus of attention (Pietrocola, 2008; Pospiech et al. 2015; Redish & Kuo, 2015; Galili, 2018; Dilling et al., 2019). Kim et al. (2018) note that physics works with real objects; therefore, the parts of the equations are connected to objects (Kim et al., 2018) and the equations themselves describe empirical (physical) situations. At this point, we see (on the epistemological level) parallels to empirically-oriented mathematics classes (Pielsticker, 2020).

7.2.1 Empirically-Oriented Mathematics Classes

This term describes mathematics classes that intentionally deal with empirical objects (real objects, Kim et al., 2018) as the mathematical objects of mathematics classes, in conception and performance (Pielsticker, 2020). Empirically-oriented mathematics classes follow a concept which, according to Schoenfeld (1985), can be described as an empirical belief system. Lessons designed in this way frequently require students to work with mathematical empirical objects (means of illustration) in math classes. Schoenfeld (1985) shows that personal belief systems matter when learning and teaching mathematics:

One's beliefs about mathematics [...] determine how one chooses to approach a problem, which techniques will be used or avoided, how long and how hard one will work on it, and so on. The belief system establishes the context within which we operate [...]. (Schoenfeld, 1985, p. 45)

Schoenfeld's (1985) characterization of the empirical belief system can be seen as an analytical tool for description and has been followed by Burscheid and Struve (2020). According to Witzke:

[t]he empirical belief system on the one hand describes a set of beliefs in which mathematics is understood as an experimental natural science, which of course includes deductive reasoning, about empirical objects. Mathematics in this sense is understood as an empirical, natural science. (Witzke, 2015, p. 307)

In this section, we will outline an empirical perspective of equations in algebra, based on the studies of Schoenfeld (1985) and Burscheid and Struve (2020). In this sense, we will use a case study to describe *what happens when students systematically tie their knowledge of equations in school algebra to empirical objects* (real objects; Kim et al., 2018).

7.2.2 Case Study – Binomial Formulas in Eighth Grade

Data collection
The data in the following case study are taken from the dissertation project of the second author of this article. For these data, eighth-grade mathematics lessons at a German middle school were observed and video recorded over a year (Pielsticker, 2020). Our example shows two students, Chris and Manuel (names changed), in math class. After the two students became familiar with and worked out the 1st and 2nd binomial formulas in math class, they wanted to justify the 3rd binomial formula.[2] Justification in this case means that the two students want to arrange the expression $(a + b) \cdot (a - b) = a^2 - b^2$ with the help of self-made

---

[2] For the case $(x + y)^2 = x^2 + 2xy + y^2$ of the binomial theorem, 3 equations in particular are learned in Germany as "kept in memory," so to speak.

1. $(a + b)^2 = a^2 + 2ab + b^2$
2. $(a - b)^2 = a^2 - 2ab + b^2$
3. $(a + b) \cdot (a - b) = a^2 - b^2$

Fig. 7.1   Tiles for the 1st binomial formula (Pielsticker, 2020)

visual aids (tiles, in the sense of "algebra tiles," Leitze & Kitt, 2000; see also NCTM[3]). With the help of 3D printing technology, the students were able to independently create tiles (Fig. 7.1) for a geometric interpretation of the binomial formulas. They had created tiles for a geometric interpretation of the 1st binomial formula (Fig. 7.1) and now wanted to use them to work out the 3rd binomial formula.

Methodology – Domains of Subjective Experience
The following excerpt from the conversation between Chris and Manuel in a mathematics lesson on the 3rd binomial formula was transcribed according to Meyer's (2010) rules. We then described and analyzed the conversation between Manuel and Chris using the concept of *Domains of Subjective Experience* (DSE), formulated by Bauersfeld (1983). This allows one to describe how students develop their knowledge in a constructivist and interactionist sense. The core idea is that learning is a domain-specific process, bound to a specific situation and context (Bauersfeld, 1983). A DSE comprises meaning, language, objects, and actions and encompasses cognitive, motor, and emotional dimensions. In this

---

[3] https://www.nctm.org/Classroom-Resources/Illuminations/Interactives/Algebra-Tiles/

article, we focus on the cognitive dimension of the DSE of two middle school students (Chris and Manuel). According to Bauersfeld:

> learning is characterized by the subjective reconstruction of social means and models through the negotiation of meaning in social interaction and in the course of related personal activities. New knowledge, then, is constituted and arises in the social interaction of members of a social group (culture), whose accomplishments reproduce as well as transmute the culture. (Bauersfeld, 1988, p. 39)

With the help of DSE, we want to describe and analyze the processes by which knowledge is developed when students systematically tie their knowledge of equations to empirical objects. According to Dilling and Witzke, it is important for our analysis that

> the descriptions of the [...] empirical objects [...] the 3D-printed models, [...] [are the] objects of reference [which] constituted the [DSEs]. (Dilling & Witzke, 2020, p. 18)

At the same time, the actions taken by the students and the terms used for the empirical objects also play a decisive role. Therefore, there is a focus on the empirical objects and the language used by the students, as we

> have not only to analyze objects that we use in mathematics classes, but we have to analyze children's associations with them, too. (Fetzer & Tiedemann, 2017, p. 1290)

We will show how the two students develop their knowledge of the 3rd binomial formula and systematically use the tiles they have created. In our example, Chris and Manuel behave as if they wanted to bring the symbolic–formal context $\left((a+b) \cdot (a-b) = a^2 - b^2\right)$ together with the context of the self-created visual aids (tiles) (Fig. 7.1).

In view of our physical interpretation in Sect. 1.3, this description is particularly interesting.

Results

The transcript excerpt (TE 1) shows how Chris and Manuel use the symbolic-formal context $\left((a+b) \cdot (a-b) = a^2 - b^2\right)$ to arrange their tiles for the 3rd binomial formula. The two students want to map every step of the formula with the help of their tiles. To clarify this situation, the transcript extract also contains descriptions of Figs. 7.2–7.7.

TE 1: Interpretation of the symbolic–formal context with the help of the tiles

	07:53	M	If we have the form here with the 2nd binomial formula, how it is here (*points to the symbolic–formal calculation*). No, look (*takes the tiles in his hand and puts them on top of each other*). It is like (*places the tiles on top of each other, as in Fig. 7.2. The largest tile is at the bottom and the other tiles are placed exactly on it*).

Fig. 7.2 Tiles on top of each other

In Fig. 7.2, Manuel puts the tiles on top of each other to represent the following equation:
$$(a + b)^2 = a^2 + 2ab + b^2$$
Manuel tries to start in this way.

08:01	M	Minus that (*takes one of the tiles for ab, removes it, and places it next to the others*).
08:05	C	And plus this one (*shows with his finger another tile for ab*).
08:07	M	And plus this one (*continues to hold the ab tiles with his fingers*)
08:09	C	Right.

Fig. 7.3 Manuel uses the symbolic–formal calculation to arrange the tiles

In Fig. 7.3, Manuel indicates the equation (symbolic–formal context), puts one tile aside, and says, "minus that" (08:01). Manuel continues doing this with other tiles he has assigned to certain terms ("And plus that" 08:07). This means he is trying to remove or add tiles according to the symbolic-formal calculation.

	08:11	C	Minus that and plus that, and then we add this one (*takes one tile in his hand, which he lays beside the paper and puts with the other tiles of the kit*). That is just those three here and not those two. Away.
**Fig. 7.4** Manuel and Chris arrange the tiles in a new way	08:15	M	Minus b^2 (*takes one tile back in his hand and arranges the tiles in a new way*).
	08:17	C	Yes but no that is not right.

In Fig. 7.4, Chris tries to arrange the tiles according to the terms $a^2 - ab + ab - b^2$ of the 3rd binomial formula. Manuel leads his finger along the symbolic–formal calculation of the formula (as if he understands the terms as individual steps) while his other hand stays on the tiles.

	08:18	M	That is the solution (*places the tiles in a certain way next to each other and one title on the tabletop, Fig. 7.5*).
**Fig. 7.5** Manuel arranges the tiles in a certain way, his solution	In Fig. 7.5, the two students place the tiles for ab, ab and b^2 in a certain way, side by side next to each other (see arrow). This representation is clearly similar to a geometric interpretation of the 3rd binomial formula. The students take the tile for b^2 away.		

| 08:21 | | M | Because minus ab, plus ab, minus b^2 *(takes away one tile and puts the same tile back and takes another tile away again, Fig. 7.6)*. |

In Fig. 7.6, Manuel interprets the steps of the symbolic-formal calculation $\left(a^2 - ab + ab - b^2\right)$ with the tiles. Each tile is assigned to a certain term. First, he puts all the tiles together. Then, he takes away the tile for ab (probably because of the minus sign). Then, he adds the token tile for ab (probably because of the plus sign). After that, he takes the tile for the term b^2 away.

Fig. 7.6 Manuel's interpretation of the symbolic–formal context with the tiles

In the transcript excerpt (TE 1), the students, Chris and Manuel, use the symbolic–formal calculation of the 3rd binomial formula to arrange their tiles correspondingly. This suggests that they are trying to map the parts of the 3rd binomial formula $\left(a^2 - ab + ab - b^2\right)$ using the tiles (TE 1, Fig. 7.6, 08:21). Chris and

Manuel interpret the terms of the 3rd binomial formula as a building plan. They take the individual terms $(a^2), (-ab), (+ab), (-b^2)$ and assign a tile to each to build the terms $(a^2 - ab + ab - b^2)$ step by step.

Interestingly, Chris and Manuel map each step with the help of their tiles (TE 1, Fig. 7.6). They create a connection to the empirical situation. The students explicitly justify every transformation (in the symbolic–formal context) using the tiles (TE 1, 08:21, Fig. 7.6). They see the terms $a^2 - ab + ab - b^2$ of the 3rd binomial formula as names corresponding to certain tiles (empirical (physical) situation), which can be added or removed according to the operation (+ or −) in front of the name. Thus, the students understand the operations as a call to action (e.g., taking away the tile for b^2) (TE 1, 08:21).

The students connect the two contexts (the symbolic–formal context $(a + b) \cdot (a - b) = a^2 - b^2$ and the context of the self-created visual aids, Fig. 7.1). The terms a^2, ab, ab, b^2 describe the empirical objects (tiles) in the concrete empirical (physical) situation (TE 1, 08:15 "minus b^2" or 08:21, "minus ab, plus ab, minus b^2"). For a normative perspective, we can include the concept of *Grundvorstellung* ("basic ideas") (vom Hofe, 1995). The following are the basic ideas of the concept of variable:

1. *object*: variable as an indeterminate object
2. *insertion*: variable as a placeholder for numbers
3. *calculation*: variable as a meaningless sign with which one operates according to certain rules (Wittmann & Malle, 1993).

Thus, we note that the work with algebra tiles focuses in particular on the aspect of "object."

7.2.3 Summary Discussion

As in the following example of physics education research (Sect. 1.3), in this case study, Chris and Manuel relate an equation (the 3rd binomial formula) to a concrete empirical (physical) situation using empirical objects. More precisely, the terms of the equation are tied to empirical objects; for the students, they are the *names* of those objects. Teachers should be aware of this when algebra is systematically linked to empirical (physical) situations in mathematics lessons.

7.3   Physics Education Research

To return to the example in the introduction (Corinne's Shibboleth), one can identify two main differences: the implicit physical interpretation of elements of the equation and the indiscriminate usage of units in mathematics. Both aspects will be analyzed in the following section. Afterwards, the main difficulties of working with equations in physics lessons will be discussed.

7.3.1   The Implicit Interpretation

Physics works with real objects; therefore, the parts of the equations are connected to those objects (Kim et al., 2018) and the equations themselves describe physical situations. There are conventions that physicists usually recall automatically when dealing with equations; for example, they associate r with a distance (usually in polar or spherical coordinates), x and y with Cartesian coordinates, and θ with an angle.

This is both a powerful and difficult way of thinking (Redish & Kuo, 2015). It is powerful because it keeps equations shorter and clearer, without mathematical conditions and background information. Therefore, equations can display a lot of information in a compressed form, in physics even more than in mathematics— if one has the necessary background knowledge (Karam et al., 2016).

At the same time, it can be difficult, as it requires a lot of hidden background knowledge and needs to be put into context. Usually, the elements of the equations are not variables, but implicit functions (Karam et al., 2016). The equation for gravitational force F, in its simplest form,

$$F = m \cdot g,$$

depends on gravitational acceleration g and the mass of an object m. For physicists, it is obvious that g depends on the mass of the source of gravitation M (usually a star or a planet), the distance to its center r, and the gravitational constant G.[4]

$$g = G \cdot \frac{M}{r^2}$$

---

[4] In addition, in physics, vector quantities—such as gravitational force F and gravitational acceleration g—are often represented in scalar form, but treated in vector form.

For mathematicians, on the other hand, this is not clear since no constraints are given. At the same time, there are implicit limits to when certain equations can be applied (Kim, 2018), especially for extremely large or small values. In this example, the given equations are no longer valid, because high values for acceleration (as occur in black holes) lead to the general theory of relativity.

The range of validity of an equation is not only limited by extreme values; often, preconditions are tacitly assumed. Redish and Kuo (2015) give the example of the photoelectric effect, in which an electron is knocked out of an atom by a photon. The photon is absorbed. There are three important physical quantities: the energy of the photon E_γ, the energy needed to expel the electron from the atom E_{Out}, and the energy E_{El} that the electron has at the end:

$$E_{El} = E_\gamma - E_{Out}$$

The energy of the photon is the available energy, some of the energy is used to knock the electron out, and the electron obtains the remaining energy as kinetic energy.

For physicists, it is obvious that this equation only applies with restrictions that are not mentioned: if the energy of the photon is less than the required energy (so $E_\gamma < E_{Out}$), the electron does not receive negative energy as the equation suggests; rather, there is no effect. Therefore, the equation is only valid when $E_\gamma > E_{Out}$. This can be represented mathematically by a Heaviside step function θ, which sets all values less than zero to zero and all values greater than or equal to zero to one:

$$\theta(x < 0) = 0$$
$$\theta(x \geq 0) = 1$$

For the photoelectric effect, the mathematically correct expression would be

$$E_{El} = \left(E_\gamma - E_{Out}\right) \cdot \theta\left(E_\gamma - E_{Out}\right).$$

In physics, these external conditions are often implied by the physical situation and not given in the equations, not even in textbooks. Physicists use their understanding of physical situations, not pure mathematics, to evaluate when to use an equation (Redish & Kuo, 2015).

To summarize, a physical equation contains more information than the pure mathematical expression suggests. Most parts of the equation have a deeper meaning. Sometimes each variable represents an entirely distinct equation. Furthermore, the context of the equation provides implicit conditions regarding when to use the equation, without specifying these conditions mathematically. From the point of view of physics, these habits have their justification in condensing physical expressions and laws, but they make it difficult for non-physicists to access the equations. The extended context of physical equations must always be considered when working with them, at least by the teacher. Ideally, background knowledge should be emphasized in the classroom. Providing a strictly mathematical extension of every equation would make them considerably more complicated in many cases. Moreover, corresponding mathematical notations are only introduced later in school or—for example, in the case of the delta distribution—not at all.

Of course, the entire background of physics—whether implicit or explicit—cannot and should not be covered in mathematics classes. However, care should be taken to ensure that physical examples used in mathematics lessons follow physical principles and do not contain elementary errors, as is the case in the following example.

7.3.2 The Use of Units

Context-oriented exercises are an important element of mathematics education. Often the context is taken from physics to show that mathematics can be applied in the (more or less) real world. However, care should be taken when including physical problems in mathematics lessons: in many cases, there are fundamental errors, even in school books. In mathematics, the context usually serves as motivation, while the core of the task is purely mathematical. Accordingly, the context is essentially stripped away after the equations have been set up, and aspects that are not necessary for the calculation are often ignored. The handling of units is often especially careless. This is apparent in the following exercise, taken from a German textbook for eighth grade (Fig. 7.7, Griesel et al., 2007; see also Karam et al., 2016):

> If an object is dropped from an aircraft at altitude h (in m) at the velocity v (in $\frac{m}{s}$), it moves approximately on a parabola with the equation $y = -\frac{5}{v^2}x^2 + h$. Here y is the height of the body and x the distance from the point of drop.
> a) An aircraft flies at speed of $6\frac{m}{s}$ and drops a supply package at an altitude of 400m. How far from the drop zone does the package land? (Griesel et al., 2007, p. 188)

An analysis of the exercise reveals that the units are wrong. The given equation should have a value in meters (or any other unit of distance) as a result, giving the altitude y, which depends on the covered distance x. While the second summand h

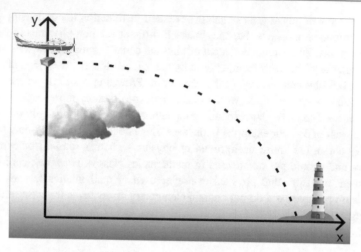

Fig. 7.7 Sketch for the exercise. (Adapted from Griesel et al., 2007, p. 188)

obviously has the unit meters, the first summand does not fit, as a simple calculation shows:

$$\left[\frac{5}{v^2}x^2\right] = \frac{1}{\frac{m^2}{s^2}} \cdot m^2 = s^2 \neq m$$

Here, meters and (squared) seconds are to be added, which makes no sense physically. This is particularly incomprehensible since a correct equation would be easy to construct: only gravitational acceleration g (see Sect. 7.3.1) with the unit $\frac{m}{s^2}$ is missing.

Incorrect handling of units not only leads to poor learning but also inhibits the application of an effective method of cognition in physics: dimensional analysis. Checking the units can help to find mistakes (cf. Sherin, 2001), but only if the exercises are constructed in a (physically) correct way. Therefore, it is not only for reasons of technical correctness that the correct use of units should be emphasized in mathematics lessons.

In fact, unit analysis is only one example of a more general aspect that is indispensable in the natural sciences: reference back to the empirical (i.e., the measurable), where units are an important component. This can be a useful and profitable addition to mathematics education and promote the link between (more theoretical) mathematics and the (largely empirical, especially in school) natural sciences.

7.3.3   The Transfer from Mathematics to Physics

In addition to the use of context-oriented tasks in mathematics teaching, the transfer from mathematics to physics is probably the greatest difficulty in using equations in physics. Students often see equations as pure tools of calculation, not (compressed) descriptions of physical phenomena (Karam & Krey, 2015). Therefore, the focus should be on understanding equations and their origins. The focus on calculation leads to the pragmatic handling of equations: often, students try to solve problems by finding a fitting equation, calculating the result, and looking for the next equation into which they can plug this result. Tuminaro and Redish (2007) call this "plug and chug," and consider it one of the most popular (but not scientific) methods to solve physical problems.

According to Karam and Krey (2015), this is mostly caused by a "vicious circle faced by physics educators" (p. 662). There is the given fact that students often use the "plug and chug" strategy to solve problems. Furthermore, Karam and Krey assume the axiom that working with equations is a learned (not intrinsic) ability, so people learn how to do it through education. Their conclusion is that students work on physical problems as they do because they have learned to do it this way. Furthermore, if these students become physics teachers, they will teach their students to solve problems in the same way (see Fig. 7.8: Vicious circle of plug and chug.).

In order to break this vicious circle, the focus in physics lessons must be on understanding the equations, not pure calculation (Karam & Krey, 2015).

This understanding is often the most difficult point in handling equations in physics. While students can handle the mathematics, their understanding of the physical situation is in many cases limited (Kim & Pak, 2002). The popular explanation that students have not yet learned the required mathematics does not stand up to closer analysis. Transferring these mathematical tools into a physical meaning is the main problem (Redish & Kuo, 2015). According to Redish and Kuo, there is one take-away message:

Fig. 7.8 Vicious circle of plug and chug

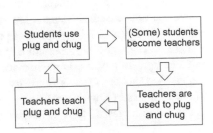

How mathematical formalism is used in the discipline of mathematics is fundamentally different from how mathematics is used in the discipline of physics – and this difference is often not obvious to students. For many of our students, it is important to explicitly help them learn to blend physical meaning with mathematical formalism (Redish & Kuo, 2015, p. 583).

Mathematical knowledge is not sufficient to be good at physics. Unlike in mathematics, both the mathematical modeling of physical situations and the physical interpretation of mathematical components are essential parts of the application of mathematics in physics. Students do not usually learn this in math class, which creates additional difficulties for them in physics class.

One way to focus more on understanding than calculation is to first try to solve a problem without any equations at all. This sets the focus on the physical meaning of the problem—which can be transferred into an equation after one grasps the physical situation. Depending on the situation, a computational step is also possible (e.g., Hohmann, 2021). Sometimes, intuition can help in this process, especially while working with graphs (e.g., potential energy diagrams, see Chapter "Functions").

Starting with an equation always involves the risk that students will believe it without understanding it. This is problematic even if they can provide a correct answer (Redish & Kuo, 2015; see Fig. 7.9), as there are significant differences in the way equations are used in mathematics and physics (Karam et al., 2016). Blind faith in a given equation quickly leads to a numerically correct result but does not promote an understanding of the connection between mathematics and physics or an understanding of physics (Fig. 7.9, left). Subsequently, the corresponding formulas can be derived from the physical solution, which also leads to the greater interconnection of the disciplines (Fig. 7.9, right).

7.4 Summary

In Sect. 1.3, the main difficulties in working with equations in physics lessons have been discussed. Students often see equations as pure calculation tools, not as (very compressed) descriptions of physical phenomena (Karam & Krey, 2015). Those descriptions, including modelling and interpretations, are an essential part of equations in physics. In mathematics classes, tasks with a physics context are often used, but the focus is generally on pure mathematics. The previously mentioned aspects of physics are largely ignored. In many cases, this leads to physically incorrect tasks in mathematics lessons. At the same time, the difference in

Fig. 7.9 Left: idealized process, starting with a calculation; Right: starting with intuition. (Original figure based on Redish & Kuo, 2015, p. 586)

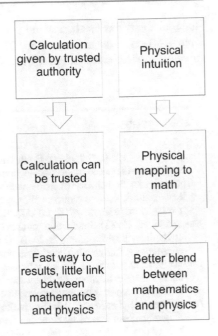

how equations are used in mathematics lessons (pure calculation) and physics lessons (modeling and interpretation) creates problems for students. It leads students to perceive calculation with physical equations as the central aspect of physics lessons, rather than the physical situations these equations describe.

Therefore, the focus should be on understanding the equations and their origins. In the case study in Sect. 1.2, we described what happens when equations (e.g., the 3rd binomial formula) themselves describe situations with empirical objects. For the students, Chris and Manuel, the terms of the equation were tied to empirical objects (tiles) and treated as their names.

These results show that students do not focus on calculating and can handle equations pragmatically. Students like Chris and Manuel understand equations with the help of concrete empirical (physical) situations. They give meaning to the symbolic–formal context by connecting it to the context of their self-created visual aids.

One way the two subjects could benefit from each other would be for mathematics and physics teachers to cooperate. By working together, physically meaningful tasks could be created, which could first be analyzed physically in physics lessons and then processed mathematically in mathematics lessons. At the same time, mathematics teachers could become accustomed to dealing with empirical (physical) contexts and situations. The associated challenges (e.g., the domain specificity of knowledge; Bauersfeld, 1988) in mathematics classes could thus be addressed.

References

Bauersfeld, H. (1983). Subjektive Erfahrungsbereiche als Grundlage einer Interaktionstheorie des Mathematiklernens und –lehrens [Subjective Fields of experience as the basis of an interaction theory of mathematics learning and teaching]. In H. Bauersfeld, H. Bussmann, & G. Krummheuer (Eds.), *Lernen und Lehren von Mathematik. Analysen zum Unterrichtshandeln II* (pp. 1–57). Köln: Aulis-Verlag Deubner.
Bauersfeld, H. (1988). Interaction, construction, and knowledge: Alternative perspectives for mathematics education. In D. A. Grouws & T. J. Cooney (Eds.), *Perspectives on research on effective mathematics teaching* (pp. 27–46). Erlbaum.
Burscheid, J., & Struve, H. (2020). *Mathematikdidaktik in Rekonstruktionen. Grundlegung von Unterrichtsinhalten [Mathematics education in reconstructions. Foundation of teaching contents].* Springer. https://doi.org/10.1007/978-3-658-29452-6.
Buschhüter, D., Spoden, C., & Borowsko, A. (2016). Mathematische Kenntnisse und Fähigkeiten von Physikstudierenden zu Studienbeginn [Mathematical knowledge and skills of physics students at the beginning of their studies]. *Zeitschrift Für Didaktik Der Naturwissenschaften, 22,* 61–75. https://doi.org/10.1007/s40573-016-0041-4.
Dilling, F., Holten, K., & Krause, E. (2019, online first). Explikation möglicher inhaltlicher Forschungsgegenstände für eine Wissenschaftskollaboration der Mathematikdidaktik und Physikdidaktik – Eine vergleichende Inhaltsanalyse aktueller deutscher Handbücher und Tagungsbände [Explication of possible research topics for a scientific collaboration in mathematics education and physics education – A comparative content analysis of current German Handbooks and conference proceedings]. *Mathematica Didactica.* From http://www.mathematica-didactica.com/Pub/md_2019/md_2019_Dilling_Holten_Krause.pdf.
Dilling, F., & Witzke, I. (2020, online first). The use of 3D-printing technology in calculus education – Concept formation processes of the concept of derivative with printed graphs of functions. *Digital Experiences in Mathematics Education.* https://doi.org/10.1007/s40751-020-00062-8.
Fetzer, M., & Tiedemann, K. (2017). Talking with objects. *CERME 10,* Feb 2017, Dublin, Ireland.
Galili, I. (2018). Physics and mathematics as interwoven disciplines in science education. *Science & Education, 27,* 7–27. https://doi.org/10.1007/s11191-018-9958-y
Griesel, H., Postel, H., & Suhr, F. (Eds.) (2007). *Elemente der Mathematik. Niedersachsen, 8. Schuljahr* [Elements of mathematics. Lower Saxony. 8th grade]. Westermann Schroedel Dieserweg. ISBN 9783507872080.
Hohmann, S. (2021). Calculating the distributions of number, mass and luminosity with the help of MS Excel. *European Journal of Physics, 42,* 015601. https://doi.org/10.1088/1361-6404/abb297.

Karam, R., & Krey, O. (2015). Quod erat demonstrandum: Understanding and explaining equations in physics teacher education. *Science & Education, 24*, 661–698. https://doi.org/10.1007/s11191-015-9743-0.

Karam, R., Uhden, O., & Höttecke, D. (2016). Das habt ihr schon in Mathe gelernt! Stimmt das wirklich? [You already learned this in Mathematics! Is this really true?]. *Unterricht Physik, 153*(154), 22–27.

Kim, E., & Pak, S.-J. (2002). Students do not overcome conceptual difficulties after solving 1000 traditional problems. *American Journal of Physics, 70*, 759–765. https://doi.org/10.1119/1.1484151.

Kim, M., Cheong, Y., & Song, J. (2018). The meanings of physics equations and physics education. *Journal of the Korean Physical Society, 73*, 145–151. https://doi.org/10.3938/jkps.73.145.

Krause, F., & Reiners-Logotheidou, A. (1978). *Kenntnisse und Fähigkeiten naturwissenschaftlich orientierter Studienanfänger in Physik und Mathematik. Die Ergebnisse des bundesweiten Studieneingangstests Physik 1978 [Knowledge and skills of science-oriented first-year students in physics and mathematics. The results of the nationwide Physics Entrance Test 1978]*. University of Bonn.

Leitze, A. R., & Kitt, N. A. (2000). Using homemade algebra tiles to develop algebra and prealgebra concepts. *Mathematics Teacher, 2000*, 462–520. https://doi.org/10.5951/MT.93.6.0462.

Marcus, S., & Watt, S. M. (2012). What is an equation? *14th International Symposium on Symbolic and Numeric Algorithms for Scientific Computing*, 23–29. https://doi.org/10.1109/SYNASC.2012.79.

Pietrocola, M. (2008). Mathematics as structural language of physical thought. In M. Vicentini & E. Sassi (Eds.), *Connecting research in physics education with teacher education (Vol. 2)*. International Commission on Physics Education. https://citeseerx.ist.psu.edu/viewdoc/download?doi=10.1.1.549.2915&rep=rep1&type=pdf.

NCTM. (2020, 8. November). *Illuminations. Algebra tiles*. https://www.nctm.org/Classroom-Resources/Illuminations/Interactives/Algebra-Tiles/.

Pielsticker, F. (2020). *Mathematische Wissensentwicklungsprozesse von Schülerinnen und Schülern Fallstudien zu empirisch-orientiertem Mathematikunterricht mit 3D-Druck [Students' mathematical knowledge development processes: Case studies of empirical mathematics education with 3D printing]*. Springer. https://doi.org/10.1007/978-3-658-29949-1.

Pospiech, G., Eylon, B., Bagno, E., Lehavi, Y., & Geyer, M.-A. (2015). The role of mathematics for physics teaching and understanding. *GIREP-MPTL 2014. Teaching/Learning Physics: Integrating Research into Practice*, 889–896. https://doi.org/10.1393/ncc/i2015-15110-6.

Redish, E., & Kuo, E. (2015). Language of physics, language of math: Disciplinary culture and dynamic epistemology. *Science & Education, 24*, 561–590. https://doi.org/10.1007/s11191-015-9749-7.

Schoenfeld, A. H. (1985). *Mathematical problem solving*. Academic.

Sherin, B. L. (2001). How students understand physics equations. *Cognition and Instruction, 19*(4), 479–541. https://doi.org/10.1207/S1532690XCI1904_3.

Tuminaro, J., & Redish, E. (2007). Elements of a cognitive model of physics problem solving: Epistemic games. *Physical Review Special Topics: Physics Education Research, 3,* 020101. https://doi.org/10.1103/PhysRevSTPER.3.020101.

Wittmann, E. C., & Malle, G. (1993). *Didaktische Probleme der elementaren Algebra [Problems of elementary algebra in mathemtics education].* Springer. https://doi.org/10.1007/978-3-322-89561-5.

Witzke, I. (2015). Different understandings of mathematics. An epistemological approach to bridge the gap between school and university mathematics. *ESU, 7,* 304–322.

Vom Hofe, R. (1995). *Grundvorstellungen mathematischer Inhalte [Basic ideas of mathematical content].* Spektrum.

Lesson Plan: Trigonometric Equations

8

Vu Dinh Phuong

Lesson Title: Trigonometric equations
Abstract: The lesson presented here deals with the topic of trigonometric equations. To motivate students, a water wheel is shown and students are asked to determine the relationship between the angle and the distance to the water. Afterwards, they graphically investigate simple trigonometric equations using function graphs. This is followed by a practice session with various given equations. Finally, students apply their new knowledge to various physics problems in the context of harmonic oscillations.

Type of school / Grade	High School / Grade 11
Prerequisites	– Basic knowledge of trigonometric functions – Basic knowledge of harmonic oscillation
Number of periods	1

V. D. Phuong (✉)
Faculty of Mathematics and Informatics, Hanoi National University of Education, Hanoi, Vietnam
e-mail: phuongvd@hnue.edu.vn

© The Author(s), under exclusive license to Springer Fachmedien Wiesbaden GmbH, part of Springer Nature 2022
F. Dilling and S. F. Kraus (eds.), *Comparison of Mathematics and Physics Education II*, MINTUS – Beiträge zur mathematisch-naturwissenschaftlichen Bildung, https://doi.org/10.1007/978-3-658-36415-1_8

Objectives	Mathematics:
	– Students understand the general equation of basic trigonometric equations: $\sin x = m$; $\cos x = m$; $\tan x = m$; $\cot x = m$
	– Students can use a calculator to calculate the approximate solutions of basic trigonometric equations
	– Students can solve trigonometric equations in simple forms
	Physics:
	Students can use trigonometric equations to solve problems related to harmonic oscillation

8.1   Methodical Commentary

The goals of the lesson are to solve simple trigonometric equations and apply them to physical problems. For this purpose, illustrations are used (e.g. the video and photos of a water wheel) as well as graphical methods (solving equations using graphs). Methodically, the students work mostly in groups, but there are also phases of individual work and whole-class discussion.

8.1.1   The Content of Equations and Inequalities in the 2018 Vietnamese Mathematics Curriculum

To better situate the lesson plan presented here, we will first give an overview of how the topic of "equations" is embedded in the curriculum in Vietnam.

- In the 2018 Vietnamese mathematics curriculum (see Table 8.1), content related to equations and inequalities is formed and developed according to a concentric and gradually expanding structure (Vietnamese Ministry of Education and Training 2018).
- Equations are taught implicitly in Grade 1 with "fill in the blank" exercises.

- From Grade 2 to Grade 7, equations are taught explicitly with "find x" exercises. These exercises are closely related to the development of arithmetic and geometry.- In Grade 8, students learn the concept of equation; specifically, they work with linear equations. In the 2018 Vietnamese mathematics curriculum, starting in Grade 8 students are asked to apply their knowledge of equations to solving integrated exercises and real life.
- In Grade 9, the content is focused on solving quadratic equations, Vieta's formulas, and linear inequalities with one variable. In addition, there is also a focus on linear equations and systems of linear equations with two variables.
- In Grade 10, students learn about linear inequalities, systems of linear inequalities with two variables, and the application of systems of linear inequalities to real-life problems. In this grade, quadratic inequalities are also introduced.
- In Grade 11, the content is focused on trigonometric equations, exponetial equations and inequalities, and logarithmic equations and inequalities. Furthermore, students are asked to apply their mathematical knowledge to solving physics problems, such as those related to harmonic oscillation and amplitude of vibration.
- The new aspect in the 2018 mathematics curriculum is the differentiation of the content in high school. From Grade 10 to Grade 12, there are three additional themes in each grade that students can choose whether or not to study. These themes relate to students' future careers. Out of the nine themes in high school, two are related to equations and inequalities: systems of linear equations with three variables in Grade 10; applying systems of linear inequalities to solving linear programming problems in Grade 12. One of the objectives of these themes is to apply mathematics knowledge to solving real-life problems and integrated problems, such as calculating resistance and current intensity in physics.

Table 8.1 The content of equations and inequalities taught in school

Topics	Grade											
	1	2	3	4	5	6	7	8	9	10	11	12
Equations, systems of equations	+	+	+	+	+	+	+	*	*	*	*	
Inequalities, systems of inequalities	+	+	+	+	+				*	*	*	

- The content taught implicitly is denoted by symbol "+"
- The content taught explicitly is denoted by symbol "*"
Source: Vietnamese Ministry of Education and Training (2019)

8.1.2 The Orientation of Teaching Equations and Inequalities in School

- Equations and inequalities are closely related to other content in mathematics. This content is extended and advanced following the development of other content. This is done in a manner appropriate to students' cognitive development. For example, the linear equation $ax+b=0$ is taught implicitly in Grade 1 and explicitly in Grade 8. However, the solutions of linear equations gradually expand from only the natural numbers to include integers, rational numbers, and real numbers. This change is based on the extension of number sets in the mathematics curriculum. In addition, solving equations in Grade 1 is very different than it is in Grade 8. In Grade 1, students are asked to choose one number that satisfies the equation without an explanation, while in Grade 8, students must solve an equation and provide an explanation of why it can be solved that way.

When functions are taught explicitly, solving equations and inequalities becomes more efficient. Questions such as "Why does this equation have no solution?" and "Why does that equation have many solutions or only one solution?" are more easily answered using functions. In the Vietnamese mathematics curriculum, the concept of function is always taught before the concept of related equation.

In the 2018 mathematics curriculum, equations are integrated with other subjects, especially physics and other STEM subjects. For example, linear equations can be applied to solving movement problems in physics, systems of linear equations with three variables can be applied to calculate resistance and current intensity in physics, and trigonometric equations can be applied to solving harmonic oscillation problems in physics.

8.2 Lesson Plan

Nr./ Time	Stage	Learning activities	Interaction form	Materials/ Resources	Methodological comments
# 1 10 min	Motivation	– Students watch part of a video of a water wheel on YouTube: https://www.youtube.com/watch?v=zGRaWTwjih4 – Students work in small groups to solve the problem: "Suppose that the distance between the horizontal axle of the water wheel and the surface of the water is 2 m. The diameter of the water wheel is 5 m. The distance between the horizontal tailrace and the surface of the water is 3.5 m. A bucket brings water from the millrace to pour it into the tail-race. Using a calculator, calculate the rotating angle of the bucket from the water surface to the tailrace."	Teacher presents/ Group work	– Video and pictures of a water wheel Source: Screenshot of YouTube video	– These activities motivate students to learn trigonometric equations – Expected answer: + $\sin a = 1.5{:}2.5 = 0.6 \Rightarrow a \approx 37^0$ + $\sin b = 2{:}2.5 = 0.8 \Rightarrow b \approx 53^0$ + The rotating angle of the bucket is approximately 90^0

Nr./ Time	Stage	Learning activities	Interaction form	Materials/ Resources	Methodological comments
# 2 15 min	Introducing new knowledge	*Activity 1. Observe the intersection points of two graphs* – Teacher divides the class into four specialized groups – The groups observe the graphs of the trigonometric functions (drawn on A4 paper) and the graph of the constant function $y = m$ (represented by the long, thin, straight stick) to recognize the general solution of the trigonometric equations: $\sin x = m$, $\cos x = m$, $\tan x = m$, $\cot x = m$ + Group 1: recognize the general solution of the equation $\sin x = m$ + Group 2: recognize the general solution of the equation $\cos x = m$ + Group 3: recognize the general solution of the equation $\tan x = m$ + Group 4: recognize the general solution of the equation $\cot x = m$ – Each group finds the answers to the following questions:	Teacher presents Group work	– Graphs of trigonometric functions drawn on A4 paper. (See Appendix 1.) – Long, straight, and thin sticks to represent the graph of a constant function	– These activities help students to recognize the general solution of basic trigonometric equations ($\sin x = m$, $\cos x = m$, $\tan x = m$, $\cot x = m$) using the graphs of the corresponding trigonometric functions *Activity 1* – The expected answers for each group: *Group 1* *Question 1* + If $m < -1$ or $m > 1$ then there is no point of intersection + If $-1 \leq m \leq 1$ then the line $y = m$ intersects the graph of the function $y = \sin x$ at many points. There are two groups of intersection points. In each group, the distance between two adjacent points is 2π *Question 2.* $x_A + x_B = \pi$ *Group 2* *Question 1* + If $m < -1$ or $m > 1$ then there is no point of intersection + If $-1 \leq m \leq 1$ then the line $y = m$ intersects the graph of the function $y = \cos x$ at many points. There are two groups of intersection points. In each group, the distance between two adjacent points is 2π *Question 2.* $x_A + x_B = 0$ *Group 3, Group 4* *Question 1.* For all values of m, the line $y = m$ intersects the graph of the function $y = \tan x$ ($y = \cot x$) at many points. The distance between two adjacent points is π *Activity 2* – The expected answers for each group: *Group 1* + If $m < -1$ or $m > 1$, then the equation $\sin x = m$ has no solution + If $-1 \leq m \leq 1$ then, the equation $\sin x = m$ has many solutions:

Nr./Time	Stage	Learning activities	Interaction form	Materials/Resources	Methodological comments
		Question 1. For what values of m does the graph of the function $y = m$ intersect the graph of function $y = \sin x$ ($\cos x$, $\tan x$, $\cot x$)? How many points of intersection do the line and the graph of the trigonometric function have? Guess the relationship between the points of intersection *Question 2.* Let A and B be the two points of intersection nearest the origin O. Determine $x_A + x_B$ or $x_A - x_B$ (if possible) – Students present the results of their discussion – Teacher comments and assesses the answers of each group *Activity 2. Find the formulas to solve trigonometric equations* – Teacher reviews that the abscissas of the intersection points are the solutions to the equations $\sin x = m$, $\cos x = m$, $\tan x = m$, $\cot x = m$			$x = x_A + 2k\pi; \; x = x_B + 2k\pi$ *Group 2* + If $m < -1$ or $m > 1$, then the equation $\cos x = m$ has no solution + If $-1 \le m \le 1$, then the equation $\cos x = m$ has many solutions: $x = x_A + 2k\pi; \; x = x_B + 2k\pi$ *Group 3* + For all values of m, the equation $\tan x = m$ has many solutions: $x = x_A + k\pi; \; x = x_B + k\pi$ *Group 4* + For all values of m, the equation $\cot x = m$ has many solutions: $x = x_A + k\pi; \; x = x_B + k\pi$ + *Teacher confirms the knowledge:* *Sin x = m* – If $m < -1$ or $m > 1$, then the equation $\sin x = m$ has no solution – If $-1 < m < 1$, then the equation $\sin x = m$ has many solutions. – Let α be one of the solutions (i.e., $\sin \alpha = m$). We have $\sin x = m \Leftrightarrow \sin x = \sin \alpha \Leftrightarrow$ $\left[\begin{array}{l} x = \alpha + 2k\pi \\ x = \pi - \alpha + 2k\pi \end{array} \right., k \in Z.$ – If $m = -1$, then $\sin x = -1$ has the solutions $x = -\frac{\pi}{2} + 2k\pi, k \in Z.$ – If $m = 1$, then $\sin x = 1$ has the solutions $x = \frac{\pi}{2} + 2k\pi, k \in Z.$ *Cos x = m* – If $m < -1$ or $m > 1$, then the equation $\cos x = m$ has no solution – If $-1 < m < 1$, then the equation $\cos x = m$ has many solutions. We have – Let α be one of the solutions (i.e., $\cos \alpha = m$). We have

Nr./Time	Stage	Learning activities	Interaction form	Materials/Resources	Methodological comments
		- Each group finds the general solution of one trigonometric equation based on the results of Activity 1 - Teacher's hint: x_A and x_B are two solutions of the equation - Each group presents their results - Teacher comments and assesses the answers of each group			$\cos x = m \Leftrightarrow \cos x = \cos \alpha \Leftrightarrow$ $\begin{bmatrix} x = \alpha + 2k\pi \\ x = -\alpha + 2k\pi \end{bmatrix}, k \in Z$ - If m=-1, then the equation cos x = -1 has the solutions $x = \pi + 2k\pi, k \in Z$ - If m=1, then the equation cos x = 1 has the solutions $x = 2k\pi, k \in Z$ $Tan\ x = m$ or $\cot x = m$ - Let α be one of the solutions (i.e., $\tan \alpha = m$ (or $\cot \alpha = m$)), then the equation $\tan x = m \Leftrightarrow \tan x = \tan \alpha$ (or $\cot x = \cot \alpha$) has the solutions: $x = \alpha + k\pi, k \in Z.$
#3 15 min	Practice	- Teacher hands out Exercise Sheet No.1 to students. Students should complete the exercises individually within five minutes - Students discuss the solutions in pairs - Teacher invites some students to write their solutions on the board - The teacher and other students comment and assess these students' results	Think/Pair/Share	+ Exercise Sheet No. 1 Exercise 1. Solve the trigonometric equations below: a) $\sin x = \frac{1}{2}$ b) $\cos x = \frac{\sqrt{3}}{2}$ c) $\tan x = 1$ d) $\cot x = \sqrt{3}$ Exercise 2. Solve the following trigonometric equations: a) $\sin x = 3$ b) $\cos x = -1$ c) $\tan x = 0$ d) $\cot x = -1$	- This activity helps students to solve concrete trigonometric equations and develops their competence - Expected solutions: Exercise 1 a) $\sin x = \frac{1}{2} \Leftrightarrow \sin x = \sin \frac{\pi}{6} \Leftrightarrow \begin{bmatrix} x = \frac{\pi}{6} + 2k\pi \\ x = \frac{5\pi}{6} + 2k\pi \end{bmatrix}, k \in Z$ b) $\begin{bmatrix} x = \frac{\pi}{6} + 2k\pi \\ x = -\frac{\pi}{6} + 2k\pi \end{bmatrix}, k \in Z$ c) $x = \frac{\pi}{4} + k\pi, k \in Z$ d) $x = \frac{\pi}{6} + k\pi, k \in Z$ Exercise 2 a) Equation has no solution b) $x = \pi + k\pi, k \in Z$

Nr./Time	Stage	Learning activities	Interaction form	Materials/Resources	Methodological comments
		• Teacher confirms important knowledge		*Exercise 3.* Solve the following equations: a) $\sin 2x = \sin 70^0$ b) $\cos (3x+1) = \cos \frac{\pi}{8}$ c) $\tan (2x-1) = \tan \frac{\pi}{5}$ d) $\cot 3x = \cot 21^0$	c) $x = k\pi,\ k \in Z$ d) $x = -\frac{\pi}{4} + k\pi,\ k \in Z$ *Exercise 3* a) $\begin{cases} x = 35^0 + k.180^0 \\ x = 55^0 + k.180^0 \end{cases},\ k \in Z$ b) $\begin{cases} x = \frac{\pi}{24} - \frac{1}{3} + \frac{2k\pi}{3} \\ x = \frac{-\pi}{24} - \frac{1}{3} + \frac{2k\pi}{3} \end{cases},\ k \in Z$ c) $x = \frac{\pi}{10} + \frac{1}{2} + \frac{k\pi}{3},\ k \in Z$ d) $x = 7^0 + k.60^0,\ k \in Z$ – *Teacher's note:* In exercises 3a and 3d, the unit on the right-hand side of the equation is degrees, so the solutions must be written in degrees
#4 5 min	Application	– The teacher asks students to study harmonic oscillation and solve some physics problems at home	Individual work	+ *Exercise sheet No. 2* *Exercise 1* The distance an object moves as a function of time can be described as follows: $x = 3 \cos (10\pi t + \frac{\pi}{2})$ where distance is measured in meters and time in seconds. At what time $t > 0$, will the object be: a) at equilibrium and moving to the right? b) at maximum amplitude? *Exercise 2.* The motion of a particle connected to a spring is described by $x = 10 \sin (\pi t)$. At what time (in seconds) is the potential energy equal to the kinetic energy?	– This activity helps students to apply their knowledge of trigonometric equations to solving physics problems

Material:

Graphs of trigonometric functions

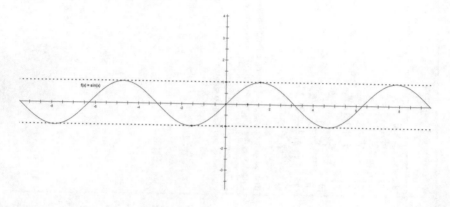

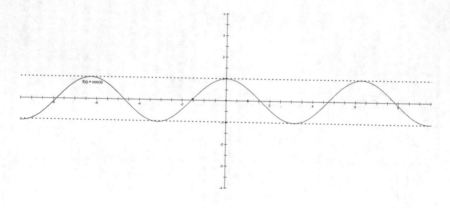

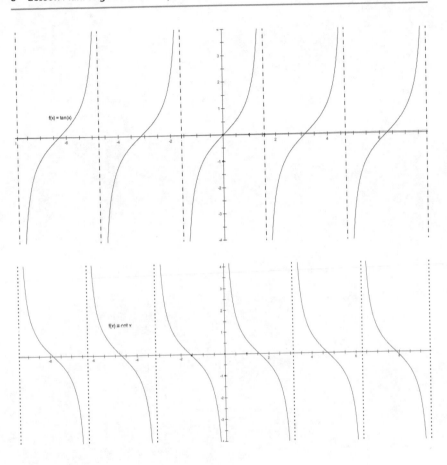

References

Vietnamese Ministry of Education and Training (2018). *The Mathematics Curriculum.*
Vietnamese Ministry of Education and Training (2019). *The training document: Studying the 2018 mathematics curriculum.*

Comparison: Functions in Mathematics and Physics Education

9

Frederik Dilling and Simon F. Kraus

9.1 Introduction

The function is one of the central concepts of mathematics, which forms the basis of the mathematical fields of calculus and analysis. A function can be defined as follows:

> A function is a relation between sets that associates every element of the first set (domain) to exactly one element of the second set (codomain). this can be notated: $f : D \to C, f \mapsto f(x)$

Functions can appear in different forms with completely different characteristics. In school, most functions considered are real functions. This means that the domain (D) and the codomain (C) represent a subset of the real numbers $\mathbb{R}$. Other types of functions that may be considered in calculus classes are sequences and series. These are functions wherein the domain (D) corresponds to the natural numbers $\mathbb{N}$.

F. Dilling (✉)
Mathematics Education, University of Siegen, Siegen, Germany
e-mail: dilling@mathematik.uni-siegen.de

S. F. Kraus
Physics Education, University of Siegen, Siegen, Germany
e-mail: kraus@physik.uni-siegen.de

F. Dilling and S. F. Kraus (eds.), *Comparison of Mathematics and Physics
Education II*, MINTUS – Beiträge zur mathematisch-naturwissenschaftlichen
Bildung, https://doi.org/10.1007/978-3-658-36415-1_9

The historical roots of the concept of function lie in geometry. The objects of investigation in differential and integral calculus were curves generated on a drawing sheet (cf. Witzke, 2009). These were embedded in a coordinate system by Descartes in the seventeenth century to introduce an algebraic description and increase the number of precisely describable curves. The word function presumably appeared the first time in a text by Leibniz from 1673, mentioned in reference to the terms constant, variable, ordinate, and abscissa. The invention of the term function is generally ascribed to Leonhard Euler, who used the term in his book *Introductio in analysin infinitorum*. He included so-called ambiguous functions, which assign several values to an initial value. The detachment from geometry occurred at the beginning of the twentieth century through the axiomatization of the foundations of mathematics by set theory, in which the concept of function was formulated as equivalent to transformation (cf. Spalt, 2019).

The development of the concept of function was essentially motivated by physical matters. In correspondence between Euler and D'Alembert beginning in 1747, the problem of the vibrating string was discussed. In this context, Euler extended the concept of function to apply differential calculus to non-static situations. Since then, the description of change with functions has been an essential part of physics.

Kjeldsen and Lützen (2015) comment on the role of physics in the development of the concept of function:

We have argued that the concept of function and its generalization to distributions gradually developed through a process that was driven by physics at crucial moments. The eighteenth century discussions about the vibrating string and the early nineteenth century study of heat conduction led mathematicians to define and gradually use functions given as a variable that depends in an arbitrary way on the value of another variable. The modern definition of function as a Cartesian product of two sets came about through an internal mathematical process of axiomatization and rigourization, but its generalization to distributions was again partly a result of physical applications. (p. 557)

In addition to its role in mathematics and physics as scientific disciplines, the concept of function plays a central role in the teaching of both subjects. In mathematics, the concept is usually introduced in middle school. Based on the knowledge of straight lines acquired in geometry, proportions are discussed first; the discussion is then expanded to the concepts of linear functions and, later, quadratic functions. In high school, differential and integral calculus are introduced for a more detailed examination of functions. Various other classes of functions are examined, including polynomial functions, trigonometric functions, exponential functions, and rational functions. References to other subjects and everyday situations play an important role (e.g., exponential functions and radioactive decay; distance and time in the context of movement). In physics classes, students learn about functions for the first time in mechanics. They are used to describe movement, so functions that depend on the parameter time are considered, particularly distance, speed, and acceleration. In the following school years, the concept of function is transferred to an increasing number of contexts and becomes a central descriptive instrument.

In Vietnam, the concept of function is officially introduced in mathematics in Grade 7 through the description of the relationship between two proportional and inverse proportional quantities. Based on the perception of the relationship between two quantities, some examples of practical models of the function are given, such as temperature–time (in a day), mass–volume (of a bar of homogenous metal), and time–velocity (in the context of uniform movement). After that, the concept of function is standardized as follows:

Ví dụ 1 : Nhiệt độ T (°C) tại các thời điểm t (giờ) trong cùng một ngày được cho trong bảng sau :

t (giờ)	0	4	8	12	16	20
T (°C)	20	18	22	26	24	21

Example 1: Temperature T (°C) at different times t (hour) of the same day, which is shown in the table below.

t (hour)	0	4	8	12	16	20
T (°C)	20	18	22	26	24	21

Ví dụ 2 : Khối lượng m (g) của một thanh kim loại đồng chất có khối lượng riêng là 7,8 g/cm³ tỉ lệ thuận với thể tích V (cm³) theo công thức : m = 7,8V.

Example 2: The mass m (g) of a homogeneous metal rod with a density of 7.8 g/cm³ is proportional to the volume V (cm³) according to the formula: $m = 7.8\,V$.

2. Khái niệm hàm số

Nếu đại lượng y phụ thuộc vào đại lượng thay đổi x sao cho với mỗi giá trị của x ta luôn xác định được *chỉ một* giá trị tương ứng của y thì y *được gọi là hàm số của* x và x *gọi là biến số.*

"If the quantity y depends on the variable quantity x such that for each value of x we always determine only one corresponding value of y then y is called a function of x and x is called the variable." (Authors' translation of the Vietnamese textbook (Than et al., 2011), Grade 7)

Then, the visual representation of the linear relationship between two quantities x and y is given on the coordinate plane. In Grade 9, the concepts of first-order functions, quadratic functions, and the increasing and decreasing properties of functions are introduced in mathematics class.

In high school, the concept of function becomes more abstract in accordance with the language of set theory. The basic concepts of functions are added: domain and codomain, even and odd properties, and variation.

1. Hàm số. Tập xác định của hàm số

Giả sử có hai đại lượng biến thiên x và y, trong đó x nhận giá trị thuộc tập số D.

Nếu với mỗi giá trị của x thuộc tập D có một và chỉ một giá trị tương ứng của y thuộc tập số thực $\mathbb{R}$ thì ta có một **hàm số.**
Ta gọi x là **biến số** và y là **hàm số** của x.
Tập hợp D được gọi là **tập xác định** của hàm số.

"Suppose there are two variable quantities x and y, where x takes value in the set of number D. If for each value of x in set D there is one and only one corresponding value of y in the set of real number $\mathbb{R}$, then we have a function. We call x a variable and y a function of x. The set D is called the domain of the function." (Authors' translation of the Vietnamese textbook (Van Hao et al., 2011), Grade 10).

Important functions in high school are cubic polynomial functions, quadratic functions, trigonometric functions, exponential functions, and logarithmic functions. The derivative is introduced in Grade 11 as an important tool to investigate the properties of functions (Chu & Nguyen, 2017).

9.2   Mathematics Education Research

9.2.1   Representational Modes of Functions

In mathematics education research, different modes of representing functions are distinguished with reference to theories on representational modes of mathematical knowledge (e.g., Bruner, 1982). In the context of functions, a distinction is often made between graphical representation (function graph), numerical representation as pairs of numbers (e.g., in a table of values), and symbolic or analytical representation as an equation:

> One of the guiding principles is the 'Rule of Three,' which says that wherever possible topics should be taught graphically and numerically, as well as analytically. The aim is to produce a course where the three points of view are balanced, and where students see each major idea from several angles. (Hughes Hallett, 1991, p. 121)

The term "Rule of Four" summarizes these three modes of representation, and additionally, so-called "verbal representation" in the form of descriptive text (see Fig. 9.1). The different modes of representation emphasize different properties of the function. For example, properties can be quickly qualitatively determined from a graph of a function. However, if a particular value, such as an extreme point, is to be determined precisely, the function equation is particularly suitable, since the method of differential calculus can be applied to it. For holistic development of knowledge, students should be able to deal with all forms of representation and switch flexibly between them. In order to achieve this, transferring representations should be explicitly addressed in class.

In school, the graphical representation of functions takes on a decisive role. In calculus teaching, concepts and relations are often derived and justified on the basis of graphs printed in textbooks or generated by digital media. The other modes of representation adopt the role of describing the graph. The students build an empirical theory for the description of graphs (cf. Witzke, 2014).

9.2.2   Conceptions of Functions

Learning a mathematical concept does not only refer to its formal definition. In order to work with a concept flexibly and in different situations, students should

Fig. 9.1 Representational modes of functions and their connections. (© Frederik Dilling)

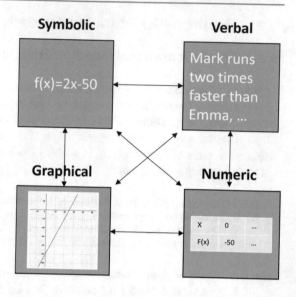

develop sustainable understandings that include different conceptions (cf. Tall & Vinner, 1981; Vom Hofe & Blum, 2016).

Therefore, one aim of calculus courses at school is the development of so-called functional thinking. This can be defined according to Vollrath (1986) as follows:

"(i) Dependences between variables can be stated, postulated, produced, and reproduced.
(ii) Assumptions about the dependence can be made, can be tested, and if necessary can be revised." (p. 387)

Therefore, functional thinking is a mental ability that enables a person to conduct certain activities. The activities summarized in i) specifically refer to the use of the concept of function, while those described in ii) are typical of mathematical thinking in general.

Vollrath (1989) clarifies the concept of functional thinking by describing three aspects regarding the concept of function that students should get to know in mathematics class. The aspect of assignment is described in the following way:

Functions are used to describe or create relationships between quantities: one quantity is then assigned to another, so that one quantity is seen as dependent on the other. (p. 8, Authors' translation)

This aspect emphasizes the unambiguous assignment of two quantities, represented in the notation $x \rightarrow y$, and also the dependence of the two quantities, represented by $y = f(x)$. The aspect of assignment is usually present in students' conception of functions, due to its connection to the definition of a function. It becomes apparent, for example, when reading a table of function values or determining function values based on a given graph. Another aspect mentioned by Vollrath (1989) is covariation:

Functions are used to determine how changes of a quantity affect a depending quantity. (p. 12, Authors' translation)

This aspect is reflected in statements such as "changing x changes y" and can be considered by reading a horizontal table horizontally or looking at the course of a graph. The aspect of covariation is often not sufficiently developed by the problems dealt with in class. Thus, various approaches have been developed to emphasize the covariation of the two parameters represented in functions. The last aspect mentioned by Vollrath (1989) is the so-called "function as a whole" aspect or object-aspect:

With functions, one considers a given or created interrelationship as a whole. (p. 15, Authors' translation)

Instead of looking at individual pairs of values, the assignment or the set of pairs of values are interpreted as a new object. This becomes particularly apparent when a function is represented as a graph since global properties such as symmetry, slope, and turning points can be identified visually. In order to consider a function as an object, it seems important to have sufficiently developed the assignment and covariation aspect.

An approach that is widely used internationally for the learning of mathematical concepts is the APOS framework (c.f. Cottrill et al., 1996). It is based on the principle that actions can be developed into repeatable processes which are encapsulated as objects and related in a broader schema. The APOS framework can be applied to the concept of function and thus describe different stages in the development of this concept.

> An action is a repeatable mental or physical manipulation of objects. [...] It is a static conception in that the subject will tend to think about it one step at a time [...]. (Dubinsky & Harel, 1992, p. 85)

An action (A) occurs as a reaction to external stimuli and is a one-step response or a series of individually triggered one-step responses. It can be a physical reflex or the recollection of facts from memory. A student whose functional conception is on the level of action may be able to insert one functional equation into another and determine special values by calculation but cannot compose two functions given in different expressions. When the student can reflect on an action and begins to establish control over it, the conception attains the level of a process (P) conception:

> A process conception of function involves a dynamic transformation of quantities according to some repeatable means that, given the same original quantity, will always produce the same transformed quantity. (Dubinsky & Harel, 1992, p. 85)

The student is in control of the transformation of the objects: he can describe and reflect on several steps without actually performing them. A process can be reversed and coordinated with other processes. Gray and Tall (1994) name this kind of conception a "procept", a neologism based on the terms process and concept. When a student is able to reflect on the transforming processes, these develop into objects (O):

> A function is conceived of as an object if it is possible to perform actions on it, in general actions that transform it. (Dubinsky & Harel, 1992, p. 85)

An object can be the result of the encapsulation of a process. When applying concepts, it is possible and often necessary to switch between an object and process view. This can be achieved by de-encapsulating the object to obtain the initial processes. This results in the development of schemas (S), which comprise actions, processes, and objects and can themselves become objects and be part of a wider schema.

9.2.3 Functions as Models for Real Phenomena

The application of mathematical knowledge to describe real phenomena is a central goal of mathematics teaching at school. In the field of functions, textbooks contain a variety of tasks in which extra-mathematical situations are described by functions. These can be divided into two types of descriptions:

1. Description of functional relationships through equations (see Fig. 9.2)
2. Description of curve progressions using function graphs (see Fig. 9.3)

Most application-oriented tasks in textbooks are not authentic modelling tasks (cf. Tran et al., 2020a). The development of a "real-world" question, its translation into mathematical language, the processing of the mathematical problem, and the "real-world" interpretation of the results usually do not take place at once. Instead, as can be seen in the two tasks above, a large part of the mathematization of the problem is already given in the tasks. In order to enable authentic modeling processes in the classroom, much more open-ended tasks and methods (e.g., heuristics, meta-cognition) must be introduced into the classroom. In par-

3   Nach starken Regenfällen im Gebirge steigt der Wasserspiegel in einem Stausee an. Die in den ersten 24 Stunden nach den Regenfällen festgestellte Zuflussgeschwindigkeit kann näherungsweise durch die Funktion f mit $f(t) = 0{,}25\,t^3 - 12\,t^2 + 144\,t$ beschrieben werden (t in Stunden, f(t) in $\frac{m^3}{h}$).
Berechnen Sie die Nullstellen von f sowie die Koordinaten der Extrempunkte von f und erläutern Sie die Bedeutung der Ergebnisse.

Fig. 9.2 Description of the water level of a reservoir with the function $f(t) - 0{,}25t^3 - 12t^2 + 144t$ (t in hours; f(t) in $\frac{m^3}{h}$). (© Brandt et al., 2015, p. 20)

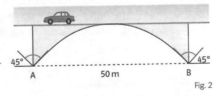

11 Ein Brückenbogen überspannt einen 50 m breiten Geländeeinschnitt. In A und B setzt der Brückenbogen senkrecht an den Böschungen auf (vgl. Fig. 2). Wählen Sie ein geeignetes Koordinatensystem, bestimmen Sie eine ganzrationale Funktion 2. Grades und berechnen Sie die Höhe des Brückenbogens.

Fig. 9.3 Description of the arc of a bridge by a quadratic function graph. (© Brandt et al., 2015, p. 33)

ticular, so-called theory-based models should be discussed: models that are not purely based on empirical observations but incorporate essential theoretical considerations. Physical phenomena are particularly suitable for this purpose (e.g., those from kinematics or nuclear physics).

9.3 Physics Education Research

Physics is often associated with the use of formulas. However, other mathematical representations usually play an even more important role in school lessons. This is especially true for functions and functional relationships, which can be represented in many ways.

9.3.1 From a Table to the Graph of a Function

First of all, it should be noted that even in the case of exploratory (i.e., apparently aimless) experimentation there is already the expectation of a result.[1] Apart from purely qualitative relationships, these expectations can be expressed in a semi-quantitative way (e.g., the greater the voltage applied to a wire, the greater the current). Such hypotheses provide the measured variables that are typically recorded in a table—the first visible step towards the exploration of a functional relationship (Barth, 2016).

It should be noted at this point that the mere observation of an experiment cannot provide quantitative information. To determine the mathematical relationship between the measured variables, further hypotheses are necessary; the quantitative relationship, therefore, does not originate directly from the measured data.[2]

The underlying question in physics lessons is that of the functional relationship between two quantities. Since a hypothesis concerning relationship has existed since the beginning, it is now necessary to examine this hypothesis critically using the data. The next step is, therefore, the visualization of the data, since trends or regularities are often difficult to identify using a table (Karam et al., 2016).

---

[1] For the methodical procedure of physics and the ideas behind it, see also Nguyen und Krause (2020).

[2] For the question of the methodical procedure in physics and the influence of mathematics therein, see also Tran et al. (2020b).

For this purpose, the loss of accuracy that is always associated with the use of diagrams compared to tabulated values is acceptable.

In contrast to mathematics lessons, in physics, there is often a choice regarding which quantity is independent and which is dependent in the graphical representation. Thus, distance–time or even velocity–time-diagrams exist in physics classes, while mathematics is usually limited to representation in the form of $f(x)$ diagrams. However, this freedom of choice influences both the representation and the interpretation that follows. For example, the characteristics of resistors can be plotted on a voltage–current or current–voltage diagram (Karam et al., 2016). Elsewhere, it is recommended to leave out the concepts of dependent and independent variables, due to the confusion in classification. Instead, the variable that is controlled by the experimenter should be plotted on the x-axis, and the y-axis contains the variable that changes as a function of x (Deacon, 1999).

The function graph has, therefore, two different purposes. On the one hand, it gives first hints of a possible function equation. On the other hand, it highlights the values that fluctuate more strongly, since these appear more clearly in a diagram than a tabular representation (Barth, 2016).

9.3.2   Diagrams and Measurement Inaccuracies

The latter point is of particular importance in the context of experimental investigations, so it will be considered separately here.

What justifies regarding measured values as "deviating" at all? Of course, measured values must first be taken seriously, as the experimental set-up provides them. However, the fundamentals of empirical science and our experience with experiments teach us that measurements are afflicted with random errors. A physical measurement without inaccuracies is inconceivable![3]

In principle, an error[4] can be distinguished from other measured values by the fact that it rarely occurs. In a classical physics lesson, however, usually, not enough measurements are recorded to enable identifying deviations by their frequency. The use of computer-assisted methods of data acquisition can help here.

---

[3] See also Volume 1, Chapter "On the Relationship between Mathematics and Physics according to Günther Ludwig" Section "Inaccuracies in the measuring process." (Geppert et al., 2020).

[4] The term "error" is used here in the sense of a single erroneous quantity and not in the sense of a failed measurement as a whole.

The question of what weight—based on the level of confidence—is given to each measured value remains insufficiently answered. However, it is crucial for the next step, in which a regression line is drawn into the diagram.

For the drawing of a regression curve or line, there is a decisive physical principle according to which, in nature, everything is "smooth" (i.e., without leaps) (Sokolowski, 2019). A regression line performs an interpolation (i.e., it compensates for values that were not measured). Thus, it represents a hypothesis, or even better, an assertion (Barth, 2016). Like any hypothesis in the natural sciences, this hypothesis cannot, in principle, be proven by the inclusion of further data. Rather, the probability of acceptance increases with the number of measured values (Nguyen & Kraus, 2020; Barth, 2016).

Occasionally, a graph is also used for extrapolation beyond the range of recorded measurement values. However, special caution is required here, since there is no experimental evidence for the validity of the initial hypothesis beyond the investigated range (Barth, 2016). This is illustrated by the following example: The deflection Δx of a spring as a function of the force F caused by the attached mass m is examined and found to be directly proportional. However, extrapolation is—contrary to the colloquial meaning of interpretation beyond the initial range—only possible within the range in which elastic deformation of the spring occurs. Outside this range, a non-reversible deformation takes place, for which the relationship $F \propto \Delta x$ no longer applies.

Another example, from the time of the transition from classical physics to quantum physics, is the so-called ultraviolet catastrophe. Based on the Rayleigh-Jeans Law, which describes the specific radiant power $B_\lambda(T)$ of a black body as a function of the wavelength λ of the light, there should be radiation tending towards infinity at short wavelengths (e.g., in the UV range). The Rayleigh-Jeans Law states

$$B_\lambda(T) = \frac{2ck_BT}{\lambda^4},$$

where c is the speed of light, k_B is the Boltzmann constant, and T is the temperature (in Kelvin). While the law accurately describes the behavior of black bodies at long wavelengths, an obvious deviation from the observations occurs, as mentioned, at short wavelengths. Also, at this point, the law—despite results in agreement with nature—can no longer be used for extrapolation. Only recourse to Planck's radiation law

$$B_\lambda(\lambda, T) = \frac{2hc^2}{\lambda^5} \frac{1}{e^{hc/(\lambda k_B T)} - 1}$$

provides the correct radiation distribution of a black body independent of wavelength. For long wavelengths, Planck's radiation law leads to the Rayleigh-Jeans Law (Passon & Grebe-Ellis, 2017).

9.3.3 From a Graph to a Formula

It has already been described that a connection is expected between the physical quantities investigated. This relationship can usually be described with a function. Representation can be provided in the form of a formula that allows for quick and easy interpolation and extrapolation.

In many cases, nature can indeed be described by simple mathematical means (i.e., relationships in the linear form $f(x) = ax + b$). However, this should not surprise us, but lead to the reflection: many physical quantities are simply defined in this way or the range of definition has been restricted accordingly (see the examples of Hook's and Rayleigh-Jeans Law) (Barth, 2016).

At this point, there is another difference between mathematics and physics: while in mathematics the notation $f(x)$ is almost omnipresent in the context of functional relationships, it is rarely found in physics. This discontinuity in representation could lead students to assume there are no functional connections in physics. Notations that are similar to the mathematical $f(x)$ can be found especially in mechanics, when it comes to the description of movements: $s(t) = \frac{1}{2}at^2 + v_0 t + s_0$ (Karam et al., 2016).

Difficulties arise when the functional relationship is not linear. If the constants are adapted accordingly, a quadratic approximation can often be adapted to the data instead of a linear one. The same applies to $\frac{1}{x^2}$ and e^{-x} relationships. Although different variants can be quickly run through with the help of a computer, choosing the most appropriate function is based on the physical interpretation. This is relevant because the slope or the point of intersection with the axis of the diagram of a straight line or curve often has a physical meaning (Barth., 2016). Table 9.1 shows some other functional relationships that occur in physics. Regardless of the problems described previously, the inspection of functional graphs is nevertheless an important step in finding functional equations, especially since corresponding statistical tests are not available for most of a student's time at school.[5]

---

[5] See also Chapter "Stochastics with a focus on probability theory".

Table 9.1 Common functions in physics and the respective meaning of slope (according to Deacon, 1999)

Experiment	Formula	x-axis	y-axis	Slope
Pendulum	$T = 2\pi\sqrt{\frac{l}{g}}$	l	T^2	$\frac{4\pi^2}{g}$
Thermal expansion of a rod	$\ell = \ell_0(1 + \alpha\Delta t)$	t	ℓ	α
Refractive index of glass	$n = \frac{\sin\Theta_i}{\sin\Theta_r}$	$\sin\Theta_r$	$\sin\Theta_i$	n
Focal length of a thin lens	$\frac{1}{f} = \frac{1}{s_1} + \frac{1}{s_2}$	$\frac{1}{s_1}$	$\frac{1}{s_2}$	-1
Discharge of a capacitor	$V_C = V_0 e^{-t/RC}$	t	$ln\ V_C$	$-\frac{1}{RC}$
Radioactive decay	$N = N_0 e^{-\lambda t}$	t	$ln\ N$	$-\lambda$

9.3.4 Functional Relations in the Physics Textbook

A look at textbooks should show which representations and combinations of variables students are typically confronted with in physics lessons. As an example, the graphic illustrations in a German textbook for Grades 9–12 have been analyzed.[6] It can be stated that the combinations of different quantities are limited in physics lessons. The frequency with which students encounter a function graph in physics class also varies.[7] The distribution is closely linked to the subject area taught (See Fig. 9.4)

A look at the independent variable, which is usually plotted on the abscissa, shows that in 41% of cases quantities depend on time t. In another 22%, quantities depend on position x or distance s. In the field of mechanics, 10% of cases depend on speed v. In other function graphs, a total of 20 further independent quantities occur, including force F, charge Q, wavelength λ, and voltage U.

In kinematics particularly, the problem of confusing a position–time graph with the path of an object is very common (Sokolowski, 2017). Obviously, it is difficult to understand that this is an abstract representation of essential quantities and not a pictorial (comic-like) "motion track."

---

[6]Although this compilation of examples does not claim to be representative, it is nevertheless not to be expected that a more extensive comparison would lead to significantly different results. The analysis was performed on a book for upper secondary school in Germany (Hoche et al., 2011).

[7]The distribution of functions among the subject areas of physics lessons will not be examined here.

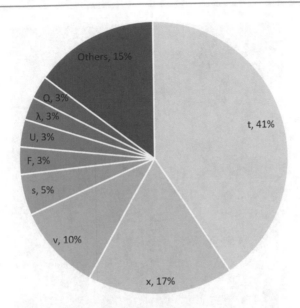

Fig. 9.4 Distribution of independent variables in a typical physics textbook

At this point, however, it should be noted that other methods of dealing with functions also exist in physics textbooks. For example, Hewitt (2015) does not use graphs to represent functional relationships. Instead, a pictorial representation is used (see Fig. 9.5). The differentiation from other forms of representation is not easy, since it contains aspects of both pictorial and graphic representation and also numerical information in the form of velocity, which is usually considered a characteristic of mathematical representations (see Pospiech, 2016).

9.3.5   Challenges in Dealing with Idealized Graphs

While idealizations are an essential part of gaining physical knowledge,[8] idealized representations—especially of functional relationships—often contain problematic elements.

---

[8] See Volume 1, Chapters "Development of Knowledge" and "Mathematization of Physics".

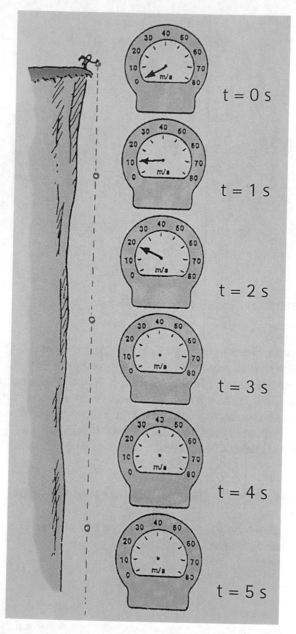

Fig. 9.5 An alternative representation of the free fall of an object. (Hewitt, 2015)

Generally, each function is represented as a curve in the Cartesian coordinate system. However, only those curves in which no vertical line intersects the curve more than once are functions (Sokolowski, 2017). This condition, also known as the vertical line test (VLT), is defined by Stewart (2001, p. 17) as follows:

A curve in XY-plane is the graph of a function of x if and only if no vertical line intersects the curve more than once.

This condition guarantees that each input value leads to exactly one unique result, which is a real number (Fig. 9.6).

Within physics lessons, however, graphs that do not pass the VLT are quite common. This is especially true for the subject area of kinematics. For a graph in which a variable such as position (or distance), velocity, or acceleration is plotted against time, passing the test means that a given object's position, velocity, or acceleration is clearly defined at all times (Sokolowski, 2017). This is exactly what is expected of a system within classical physics. However, such graphs, despite their prevalence, are often afflicted with errors resulting from excessive idealization.

If a graph does not pass the test, its rate of change can no longer be determined meaningfully because it is not continuous. Once again, the previously mentioned physical principle comes into play, which states that nature behaves continuously (Leibniz's principle of continuity). Such a graph, or rather its gradient, therefore no longer has any physical meaning (Sokolowski, 2017).

Sometimes, to avoid obvious problems, the drawing of vertical components is omitted (e.g., in the representation of the deceleration process of a vehicle; see Fig. 9.7). Further investigation is therefore necessary. Sokolowski (2019)

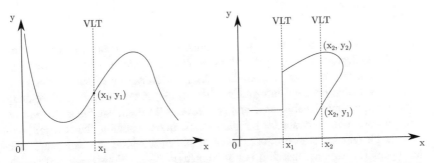

Fig. 9.6 Two examples for the investigation of graphs using the vertical line test. While Figure A is a function graph, Figure B does not pass the VLT (based on Sokolowski, 2017)

Fig. 9.7 Unrealistic
velocity–time graph

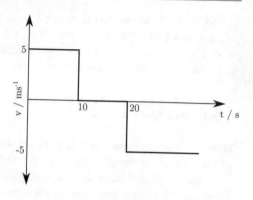

summarizes the prerequisites for graphs that correctly represent a physical motion
sequence stating that they must:

1. represent an algebraic function
2. be continuous
3. be differentiable.

Condition (1) is covered by the VLT, but this alone does not guarantee that the
representation of the movement is realistic. To fulfill Condition (2), regarding
continuity, a graph—from a purely mathematical point of view—must fulfill three
further conditions.

1. $f(x)$ is defined at $x = a$
2. $\lim\limits_{x \to a^+} f(x) = \lim\limits_{x \to a^-} f(x)$
3. $\lim\limits_{x \to a} f(x) = f(a)$

At first glance, a continuous graph may appear to be an accurate representation
of a physical process. Figure 9.8 shows a possible velocity–time diagram. As
previously noted, however, the gradient (i.e., the derivation of such a diagram)
contains valuable physical information. The graph of the derivation of velocity
with regard to time—that is, acceleration—reveals the dilemma. Figure 9.8 shows
two possible solution strategies for such a task. In one solution, the acceleration–
time graph remains close to the original graph. In this case, the new graph will
have gaps (i.e., the acceleration is not defined at the positions $t = a$ and $t = b$).

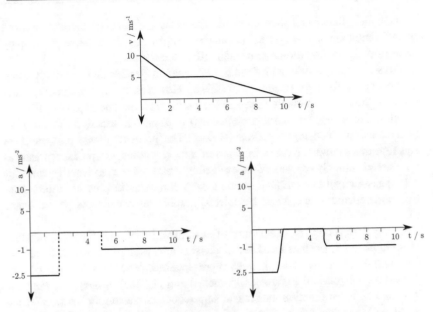

Fig. 9.8 A non-differentiable velocity–time diagram and two conceivable solutions for corresponding acceleration-time diagrams (based on Sokolowski, 2019)

Another solution maintains differentiability by creating "smooth" transitions at the critical points. However, this may lead to larger deviations so that the underlying motion is no longer accurately reproduced.

Accordingly, if a task is to be solvable, the starting material must also meet Condition (3), differentiability, which includes the additional condition:

$$\lim_{x \to a^+} f'(x) = \lim_{x \to a^-} f'(x) = f'(a).$$

9.4   Conclusion and Outlook

A glance at history shows that the development of the concept of function was essentially caused by physical problems. It can also be seen that the roots of the function are found in the field of geometry. Today the concept of function, due to the axiomatization of mathematics, is defined by set theory. In school mathematics, linear and then quadratic functions are introduced first, followed by differentiation and integration in calculus. In physics education, as well, many kinds of functional relations are used in different contexts.

Physics usually works with formulas or equations[9] rather than explicitly using the concept of function (Karam et al., 2016). However, in most textbooks, functions (in the form of function graphs) are very common. The objective of using functions (or their graphical representations) in physics is usually to find a functional relationship using experimental data. The product of this process is an equation that allows for the calculation of new quantities. Graphical representations are also highly relevant when evaluating measured data to identify unavoidable measurement errors. Thus, dealing with functions in physics differs little from mathematics since, in both subjects, graphic representations play a major role.

In physics classes, representations of movement are often shown as a function of time, position, or distance. Even if physics always depends on idealizations, this should not be excessive in such representations. The consideration of fundamental mathematical (unique assignments) and physical (principle of continuity) conditions for function graphs can help reduce confusion for learners. When mathematics makes use of functional connections, the chosen tasks are often not authentic[10] (cf. Dilling & Krause, 2020). Authentic modeling usually does not take place, either, since the central elements of mathematization are already given, so an opportunity for beneficial learning is missed.[11]

In contrast to mathematics instruction, functions in physics are not considered from the perspective of set theory. In addition to the table, the representation of the function graph plays a major role. In fact, drawing the graph of a function can itself be regarded as a method (or tool) of physics when it comes to error evaluation or the identification of functional relationships. At this point, intuition and experience usually play a more important role than mathematical methods. In both subjects, the use of authentic data offers the potential to show the limits of functional relationships, which often result from the fact that the material is only based on functions that have already been discussed in class.

---

[9] See Chapter "Equations in Mathematics and Physics Education".

[10] See Volume 1, Chapter "Interdisciplinarity in School and Teacher Training Programs" (Nguyen & Krause, 2020).

[11] See Volume 1, Chapter "Models and Modeling" (Tran et al., 2020a).

References

Barth, M. (2016). Tabelle - Graph - Formel... und zurück Ein knapper Überblick [Table - Graph - Formula... and back A brief overview]. Naturwissenschaften Im Unterricht. Physik, 27(153/154), 56–57.

Brandt, D., Jörgens, T., Jürgensen-Engl, T., Riemer, W., Schmitt-Hartmann, R., Sonntag, R., & Spielmans, H. (2015). *Lambacher Schweizer Mathematik Qualifikationsphase. Leistungskurs/Grundkurs*. Nordrhein-Westfalen, Stuttgart, Leipzig: Ernst Klett.

Bruner, J. S. (1982). *Toward a theory of instruction*. Harvard University Press.

Chu, C. T., & Nguyen, T. D. (2017). Analysis of didactic transposition in teaching the concept of derivative in high schools in the case of Vietnamese textbooks in 2000, 2006 and American textbook in 2010. *HNUE Journal of Science, Educational Science, 62*(6), 10–18. https://doi.org/10.18173/2354-1075.2017-0123

Cottrill, J., Dubinsky, E., Nichols, D., Schwingendorf, K., Thomas, K., & Vidakovic, D. (1996). Understanding the limit concept: Beginning with a coordinated process scheme. *Journal of Mathematical Behavior, 15*, 167–192.

Deacon, C. (1999). The importance of graphs in undergraduate physics. *The Physics Teacher, 37*(5), 270–274. https://doi.org/10.1119/1.880285

Dilling, F., & Krause, E. (2020). Zur Authentizität kinematischer Zusammenhänge in der Differentialrechnung – Eine Analyse ausgewählter Aufgaben [On the authenticity of kinematic relations in differential calculus – An analysis of selected exercises]. *MNU-Journal, 2*(2020), 163–168.

Dilling, F., Stoffels, G., & Witzke, I. (2020). (in print). Springer Spektrum: Didaktik der Analysis.

Dubinsky, E., & Harel, G. (1992). The process conception of function. In G. Harel & E. Dubinsky (Eds.), *The concept of function: Aspects of epistemology and pedagogy* (pp. 85–106). Mathematical Association of America.

Geppert, J., Krause, E., Nguyen, P. C., & Tran, N. C. (2020). On the Relationship between Mathematics and Physics according to Günther Ludwig. In I. Witzke & O. Schwarz (Series Eds.) & S. F. Kraus & E. Krause (Vol. Eds.), *Reihe der MINT-Didaktiken der Universität Siegen. Comparison of Mathematics and Physics Education I: Theoretical Foundation for Interdisciplinary Collaboration* (pp. 137–156). Springer. https://doi.org/10.1007/978-3-658-29880-7_8

Gray, E. M., & Tall, D. (1994). Duality, ambiguity and flexibility: A proceptual view of simple arithmetic. *Journal of Research in Mathematics Education, 26*, 115–141.

Hewitt, P. G. (2015). *Conceptual physics. Always learning*, (12. ed., global ed.). Pearson.

Hoche, D., Küblbeck, J., Meyer, L., Reichwald, R., Schmidt, G.-D., Schwarz, O., & Spitz, C. (2011). *Duden Lehrbuch Physik – gymnasiale Oberstufe* (2., vollst. überarb. Ausg). Berlin: Duden Schulbuch.

Hughes Hallett, D. (1991). Visualization and calculus reform. In W. Zimmermann & S. Cunningham (Eds.), *Visualization in teaching and learning mathematics, MAA Notes No. 19*, p. 121–126.

Karam, R., Uhden, O., & Höttecke, D. (2016). Das habt ihr schon im Mathe gelernt! Stimmt das wirklich?: Ein Vergleich zwischen dem Umgang mit mathematischen Konzepten in der Mathematik und in der Physik [You already learned that in math! Is it really true?: A comparison between the use of mathematical concepts in mathematics and physics] Naturwissenschaften Im Unterricht. *Physik, 27*(153/154), 22–27.

Kjeldsen, T. H., & Lützen, J. (2015). Interactions between mathematics and physics: The history of the concept of function—Teaching with and about nature of mathematics. *Science & Education, 24*(5–6), 543–559.

Nguyen, P. C., & Krause, E. (2020). Interdisciplinary in school and teacher training programs. In S. F. Kraus & E. Krause (Eds.), *Comparison of mathematics and physics education I: Theoretical foundations for interdisciplinary collaboration* (pp. 15–35). Springer Fachmedien Wiesbaden. https://doi.org/10.1007/978-3-658-29880-7_2

Nguyen, V. B., & Kraus, S. F. (2020). The Nature of science. In I. Witzke & O. Schwarz (Series Eds.) & E. Krause & S. F. Kraus (Vol. Eds.), *Reihe der MINT-Didaktiken der Universität Siegen. Comparison of mathematics and physics education I: Theoretical foundation for interdisciplinary collaboration*. Springer.

Passon, O., & Grebe-Ellis, J. (2017). Planck's radiation law, the light quantum, and the prehistory of indistinguishability in the teaching of quantum mechanics. *European Journal of Physics, 38*(3), 35404. https://doi.org/10.1088/1361-6404/aa6134.

Pospiech, G. (2016). Formeln, Tabellen und Diagramme. Einsatz verschiedener mathematischer Darstellungsformen im Physikunterricht [Formulas, tables and diagrams. Using different forms of mathematical representation in physics lessons]. Naturwissenschaften Im Unterricht. *Physik, 27*(153/154), 14–21.

Sokolowski, A. (2017). Graphs in kinematics—A need for adherence to principles of algebraic functions. *Physics Education, 52*(6), 65017. https://doi.org/10.1088/1361-6552/aa873d

Sokolowski, A. (2019). Graphs in physics—A need for adherence to principles of function continuity and differentiability. *Physics Education, 54*(5), 55027. https://doi.org/10.1088/1361-6552/ab2943

Spalt, D. D. (2019). *Eine kurze Geschichte der Analysis für Mathematiker und Philosophen [A short history of calculus for mathematicians and philosophers]*. Springer Spektrum.

Tall, D., & Vinner, S. (1981). Concept image and concept definition in mathematics with particular reference to limits and continuity. *Educational Studies in Mathematics, 12*(2), 151–169.

Than, T., Binh, V. H., & Duc, P. G. (2011). *Mathematic Textbook Grade 7* (8th Edition). Vietnam Education Publishing House.

Tran, N. C., Chu, C. T., Holten, K., & Bernshausen, H. (2020a) Models and modeling. In: S. Friedrich Kraus & E. Krause (Eds.), *Comparison of mathematics and physics education I. Theoretical foundations for interdisciplinary collaboration* (S. 257–298). Springer Spektrum (MINTUS – Beiträge zur mathematisch-naturwissenschaftlichen Bildung (MINTBMNB)).

Tran, N. C., Nguyen, P. C., Krause, E., & Kraus, S. F. (2020b). The Mathematization of physics throughout history. In I. Witzke & O. Schwarz (Series Eds.) & E. Krause & S. F. Kraus (Vol. Eds.), *Reihe der MINT-Didaktiken der Universität Siegen. Comparison of mathematics and physics education I: Theoretical foundation for interdisciplinary collaboration*. Springer.

Van Hao, T., Tuan, V. & Cuong, D. M. (2011). *Mathematic Textbook Grade 10* (5th Edition). Vietnam Education Publishing House.

Vollrath, H.-J. (1986). Search strategies as indicators of functional thinking. *Educational Studies in Mathematics, 17*(4), 387–400.

Vollrath, H.-J. (1989). Funktionales Denken [Functional thinking]. *Journal Für Mathematik-Didaktik, 10*(1), 3–37.

Vom Hofe, R., & Blum, W. (2016). "Grundvorstellungen" as a category of subject-Matter didactics. *Journal Für Mathematik-Didaktik, 37*(1), 225–254.

Witzke, I. (2009). *Die Entwicklung des Leibnizschen Calculus: Eine Fallstudie zur Theorieentwicklung in der Mathematik [The Development of Leibniz's Calculus: A Case Study in Theory Development in Mathematics]*. Franzbecker.

Witzke, I. (2014). Zur Problematik der empirisch-gegenständlichen Analysis des Mathematikunterrichtes [On the problem of empirical-objective analysis of mathematics teaching]. *Der Mathematikunterricht, 60*(2), 19–31.

Lesson Plan: Quadratic Functions— Graphs and Applications

10

Chu Cam Tho

Lesson Title: Quadratic functions: graphs and applications
Abstract: The topic of the lessons presented here is the quadratic function. Students learn about the properties of quadratic functions and apply them later in the lesson to real-world problems, such as the height of a bridge and the shape of a water fountain. In this way, students acquire both mathematics and physics skills.

Type of school / Grade	High School / Grade 10
Number of periods	2

C. C. Tho (✉)
Research Division On Educational Assessment (RDEA),
The Viet Nam Institute of Educational Sciences (VNIES), Hanoi, Vietnam

© The Author(s), under exclusive license to Springer Fachmedien Wiesbaden
GmbH, part of Springer Nature 2022
F. Dilling and S. F. Kraus (eds.), *Comparison of Mathematics and Physics
Education II*, MINTUS – Beiträge zur mathematisch-naturwissenschaftlichen
Bildung, https://doi.org/10.1007/978-3-658-36415-1_10

151

Objectives	Mathematics: – Students can make a table of values for a quadratic function – Students can graph quadratic functions – Students can recognize the basic characteristics of a parabola, such as its vertex and axis of symmetry – Students can identify and interpret the properties of quadratic functions from a graph – Students can apply knowledge of quadratic functions and graphs in solving real-world problems Physics: – Students can interpret the formula for average speed to define speed in one direction – Students can apply the formula to calculate speed and acceleration

10.1 Methodical Commentary

The goal of this lesson is to learn about quadratic functions and their application to real-world problems. For this purpose, a variety of tools are used, such as worksheets, the Internet, and the computer software program GeoGebra. In addition, students complete tasks in different social combinations, with a focus on group work.

10.2 Lesson Plan

No./ Time	Stage	Learning activities	Interaction form	Materials/ Resources	Methodological comments
# 1 10 min	Motivation	+ Students observe the pictures of the trajectories of some objects and predict which function is suitable for each trajectory + Students determine quadratic functions based on graphs using GeoGebra software + Teacher identifies the general form of a quadratic function	+ Teacher presents + Group work	+ Pictures of the trajectories of objects with the form of a parabola + Computer/ phone with GeoGebra pre-installed	+ Inquiry-based learning + Some limitations: Not enough devices, lack of ICT skills when using GeoGebra software + Students can investigate the problem after the lesson as homework
# 2 25 min	Introducing new knowledge	+ In pairs, students discuss the conditions that make a function with the form $y = ax^2 + bx + c$ a quadratic function + Students work in groups to investigate some properties of quadratic functions based on the coefficients a, b, c + Students present group work. Teacher corrects students' answers + Students complete the exercises on Worksheet 1	+ Group work + Teacher presenting	+ Pre-printed learning worksheet	+ Think – pair – share T (Think): The teacher begins by asking specific questions: When does the concave graph open up/down? – What is the formula to calculate the coordinates of the vertex of a quadratic function? – What are the variations of the function (increasing, decreasing properties)? – What is the axis of symmetry of the function? – Where does the graph of a quadratic function intersect the y-axis? Students "think" about what they know or have learned about each topic P (Pair): Each student is paired with another student or a small group S (Share): Students share their thinking with their partners. The teacher expands the "share" into a whole-class discussion + Tablecloth strategy

No./ Time	Stage	Learning activities	Interaction form	Materials/ Resources	Methodological comments
# 3 10 min	Practice	+ Students complete the exercises on Worksheet 2	Individual work Group work	Pre-printed learning worksheet	+ Students work in pairs with support from the teacher + Students may find it difficult to complete Exercise #3 because the information is implicit

Worksheets:

Worksheet 1

Q2.1 Which of the following functions is a quadratic function? identify the values of a, b, and c for the quadratic functions.

a) $y = x^2 - 4x + 3$	b) $y = 6 + 5x - x^2$	c) $y = (2x + 1)^2$
d) $y = -x^2 - 9$	e) $y = ax^2 (a \neq 0)$	g) $y = (3x + 1)^2 - (3x - 2)(3x + 2)$

Q2.2 For what values of m are the following quadratic functions?

a) $y = (3m - 5)x^2 + x - 1$	b) $y = (m^2 - 7m + 6)x^2 + 9$

Worksheet 2

Exercise 1: Without graphing, define the vertices, axis of symmetry, concave direction, table of variation, and maximum/minimum values of each function:

a) $y = x^2 - 4x - 1$	b) $y = -x^2 - 2x - 2$	c) $y = -x^2 + 4x$

Exercise 2: Which function is in accordance with the graph below? Explain.

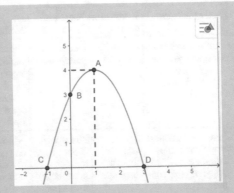

Exercise 3: Determine the sign of the coefficients a, b, c of $y = f(x) = ax^2 + bx + c(a \neq 0)$ in each of the following cases:

a)   b)

c)   d)

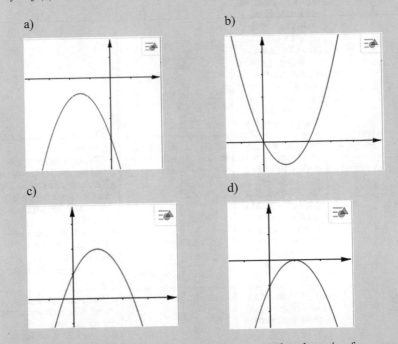

Exercise 4: Find the value of the parameter m so that the axis of symmetry of the graph of $y = x^2 - (m + 2)x + 3m - 2$ passes through the point $M(4, 3)$.

Nr./ Time	Stage	Learning activities	Interaction form	Materials/ Resources	Methodological comments
# 5 5 min	Motivation	+ Students express their opinions about the application of quadratic functions to real life	Group work	Online computer/phone	Brainstorming activity: In rapid ideation, everyone writes down as many ideas as possible in a set amount of time before any ideas are discussed, critiqued, or fleshed out
# 6 5 min	Introducing new knowledge	+ Teacher introduces the applications of quadratic functions to physics and construction + In Physics 10, the relationship between distance and time in uniform accelerated motion (increasing or decreasing) is represented by a quadratic function + Teacher rewrites the formula for calculating distance s in terms of time t and acceleration a $s = \frac{1}{2}at^2 + v_0 t + s_0$	Teacher asks, students answer	Slideshow about relevant knowledge from Grade 10 Physics	+ Students should focus on the conditions in which one can apply the formula + What do the letters s, a, and v stand for?
# 7 5 min	Practice	An object starts in uniform accelerated motion (increasing) in a straight line with an initial speed of zero. After traveling 100 m, the object reaches a speed of 10 m/s a) Calculate the acceleration of the object	Group work	Slideshow of practice examples	Think-pair-share T (Think): Teachers begin by asking specific questions: • How can we calculate the acceleration of an object with the initial information given? • Which formula can we apply in this situation? Students "think" about what they know about the topic
		b) Write an expression for the distance the object has travelled over time + In addition, quadratic functions have a number of other applications to real life			P (Pair): Each student is paired with another student or a small group S (Share): Students share` their thinking with their partners. The teacher expands the "share" into a whole-class discussion

Nr./ Time	Stage	Learning activities	Interaction form	Materials/ Resources	Methodological comments
# 8 30 min	Application	Students complete the exercises on Worksheet 3	Group work	Pre-printed learning worksheet	+ Tablecloth strategy: Students present their work on A3 paper divided into two sections. The outside area shows individual work; the center area shows the conclusions of the group after discussion + Students may find it difficult to determine the appropriate coordinates when sketching the graph of a quadratic function

Worksheets:

Worksheet 3
Problem 1: The height of Truong Tien Bridge

Truong Tien Bridge, also known as Thanh Thai Bridge, Nguyen Hoang Bridge, Clémenceau Bridge and Trang Tien Bridge, is a ring-shaped bridge spanning the Perfume River. This was one of the first bridges with a steel structure built in Indochina in the late 19th and early twentieth centuries with new Western techniques and materials.

The bridge was designed in the style of Gothic architecture by the famous architect Gustave Eiffel, who created the Eiffel Tower (symbol of France) and the Statue of Liberty (symbol of America).

Truong Tien Bridge has a length of 402.6 m and a width of 6 m. It has six spans of conical-shaped steel girder, each of which has a length of 67 m.

Image credits: left: author: ntt, license:CC BY-SA 30, https://creativecommons.org/licenses/by-sa/30/legalcode; right: image detail, author: Supanut Arunoprayote, license: CC BY40, https://creativecommons.org/licenses/by/40/legalcode

Calculate the height of Truong Tien Bridge (i.e., the height of each bridge span). You know that a vertical rope hanging from a position on the bridge span 3 m above the surface of the bridge touches the surface of the bridge at a point 11 m from the nearest bridge span.

Problem 2:
A firefighter stands $d = 20$ m from a burning building, spraying water from a hose at an initial angle of $\theta_i = 45°$ from the ground. The initial speed of the water flow is $v_0 = 20$ m/s. Determine the height of the burnt part of the building.

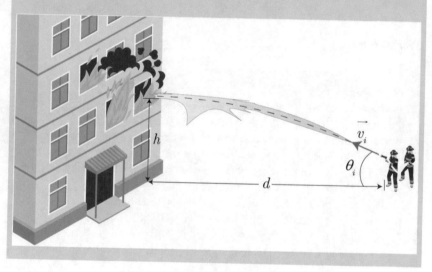

Solutions for Worksheet 3

Problem 1:

Each span mirrors the graph of the quadratic function $y = ax^2 + bx + c\,(a \neq 0)$.

We can orient the x-axis vertically, with one foot of the span at the origin and the upward direction being positive. The other foot of the span is placed at $A(67, 0)$. The height of the span is the y–coordinate of the vertex of the parabola.

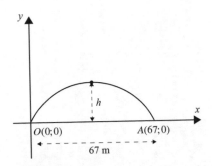

Because $O(0; 0)$; $A(67; 0)$; $M(11; 3)$ lie on the graph of (P), we can deduce:

$$\begin{cases} c = 0 \\ 67^2 a + 67b = 0 \\ 11^2 a + 11b = 3 \end{cases} \Leftrightarrow \begin{cases} c = 0 \\ a = \frac{-3}{616} \\ b = \frac{201}{616} \end{cases}$$

So, the height of one span of Truong Tien bridge is $h = -\frac{\Delta}{4a} \approx 4,4655$ m.

Problem 2:
+The motion trajectory of an oblique throw is a parabola described by the equation:

$$y = -\frac{g}{2v_0^2 \cdot cos^2\alpha} x^2 + tan\, \alpha \cdot x$$

In which: g: acceleration of gravity (equal to 9.8 m/s^2).

* v_0: initial velocity
* α: initial throw angle (relative to the horizontal axis)

+Applying this equation to the problem, the equation describing the flow of the water is: $y = -\frac{9,8}{2.20^2 \cdot cos^2 45°} x^2 + tan\, 45° \cdot xy = -\frac{9,8}{2.20^2 \cdot cos^2 45°} x^2 + tan\, 45° \cdot x$.
    Assuming that the fire is h (m) above the ground, this position corresponds to the point $A(20; h)A(20; h)$ on the coordinate plane. We can deduce that $h = -\frac{9,8}{2.20^2 \cdot cos^2 45°} 20^2 + tan\, 45 \cdot 20 \Leftrightarrow h = 10.2(m)h = -\frac{9,8}{2.20^2 \cdot cos^2 45°} 20^2 + tan\, 45° \cdot 20 \Leftrightarrow h = 10,2(m)$.
The height of the burnt part of the building is 10.2 m.

Lesson Plan: Projectile Motion

11

Tuong Duy Hai

Lesson Title: Projectile motion
Abstract: In these two lessons, students learn how to determine the trajectory and maximum range of a projectile that is subject to the acceleration of gravity. The starting point is real-life problems from sports in which objects have to be thrown or shot as far as possible. The lessons are further enriched with experiments to support the development of the kinematic equations to describe trajectories and further motivate students. Video analysis software is also used to analyze the trajectory of horizontal throws. The result of the optimal throwing angle, which is derived through mathematical means and tested experimentally, is ultimately transferred back to real-life situations in sports.

Type of school / Grade	High School / Grade 10
Prerequisites	Knowledge of the concepts and properties of – parabolas – uniformly accelerated rectilinear motion
Number of periods	2

T. D. Hai (✉)
Hanoi National University of Education, Faculty of Physics, Hanoi, Vietnam
e-mail: haitd@hnue.edu.vn

© The Author(s), under exclusive license to Springer Fachmedien Wiesbaden
GmbH, part of Springer Nature 2022
F. Dilling and S. F. Kraus (eds.), *Comparison of Mathematics and Physics
Education II*, MINTUS – Beiträge zur mathematisch-naturwissenschaftlichen
Bildung, https://doi.org/10.1007/978-3-658-36415-1_11

Objectives	Mathematics – Students identify that the trajectory of a projectile in the Cartesian coordinate system is a parabola, determine the coordinates of its peak, and the solution to a 2nd-order equation Physics – Students identify that the trajectory of a projectile in the Cartesian coordinate system is a parabola, determine the coordinates of its peak, and the solution to a 2nd-order equation

11.1 Methodical Commentary

The teaching sequence is designed based on the following five questions and considerations:

1. Initial question: How does one throw an object as far as possible?
2. Main question: What factors does the range of a projectile depend on?
3. Hypothesis: The range of a projectile depends on its initial speed and initial angle.
4. Result: The formula to determine the range of a projectile.
5. Conclusion: Throwing an object at a 45-degree angle gives maximum range. The trajectory of a projectile is a parabola and its range depends on the initial speed and angle.

In the following sections, some methodological specifics of the individual phases of the two lessons will be discussed.

Phase 1

While the video plays, the teacher introduces a series of terms to the students: starting point, ending point, range, the height of an object. Then, students freely discuss how to throw an object as far as possible. If students cannot predict the factors that characterize a projectile's motion (initial velocity, initial angle), the teacher should make suggestions (how to throw an object in the air; do the range and maximum height of a projectile change when the initial throwing conditions change?). Students are then asked to present their methods for throwing an object. Next, the teacher suggests that to determine which method is best, they need to determine the factors on which the range of a projectile depends. In order to determine the range, the trajectory must also be determined.

Phase 2

As an introductory experiment in Phase 2, throw a tennis ball parallel to the board in order to describe the shape of its trajectory. Before throwing the tennis ball, the teacher asks students to predict its trajectory (Is it curved or straight? Is it curved upward or downward? What forces are acting on the ball as it moves?). As students discuss, give them a hint about the similarity between the trajectory of a projectile and that of water sprayed by a sprinkler. The water drops have the same initial velocity and angle, so their trajectory looks like that of a projectile. Ask one or two groups to present their picture of the trajectory and predict its shape. The teacher then highlights the starting point, ending point, and symmetry, draws the trajectory on the board, and identifies its shape as a parabola.

Phase 3

The teacher draws the predicted trajectory on the board and directs students to choose the appropriate coordinates and draw the vectors of initial velocity and acceleration. The teacher then asks students to work on Worksheet 1. While the students are working and discussing as a group, the teacher should review the kinematic equations for motion with constant acceleration if students have difficulty with them. Observe the groups and ask them to remove the time variable from the equation, find the connection between x and y in the form: $y = ax^2 + bx + c$, and then analyze the coefficients with reference to the properties of a parabola (e.g., second-order coefficient is negative, there is no coefficient c, parabolic curve, etc.).

Phase 5

The teacher identifies that the trajectory of the projectile is a parabola that corresponds to the kinematic equation of motion derived in Phase 3. Ask the students to determine the range based on the following clue: The final horizontal displacement of the projectile is twice that of the vertex because of the symmetry of the parabola. Suggest that students use trigonometry to deduce that the expression for the range depends on the square of the initial velocity and the sine of angle 2α.

Phase 6

The teacher describes the experiment: how to set the angle, how to create a constant initial speed using a bottle with a tube inserted into the cap and nearly reaching the bottom of the bottle, and how to measure the range. The teacher creates a table with different angles (30, 40, 45, 50, 60) and their corresponding ranges. One can conclude from the table that the range reaches its maximum at $45°$. Use this to confirm the derived expression.

11.2 Lesson Plan

No./Time	Stage	Learning activities	Interaction form	Materials/ Resources	Methodological comments
# 1 10 min	Creating real-life situation	Present a video introducing several sports: javelin throw, shot put, basketball, volleyball, football. The teacher writes the question on the blackboard: "How does one throw an object as far as possible?"	Watch videos individually, then discuss in groups under the guidance of the teacher	Video clips, projector, laptop, blackboard, paper	– Introduction of new terms (starting point, ending point, range, height of an object) – Determination of the trajectory as a method to determine the maximum throwing distance

No./Time	Stage	Learning activities	Interaction form	Materials/ Resources	Methodological comments
# 2 10 min	Prediction for the trajectory of a projectile	The teacher carries out an experiment: throw a tennis ball parallel to the board. Then the students are asked to describe the trajectory of the ball After that, the teacher presents stroboscopic photos or a slow-motion video of a projectile to determine its trajectory on the blackboard	Each individual plots the position of the object on paper, then groups discuss its trajectory	Tennis ball, blackboard, video, video analysis software, projector, laptop	– Collect trajectory predictions in advance – During the discussion, point out the similarity between the trajectory of a projectile and water from a sprinkler – One or two groups present their trajectories, emphasizing the peak, initial point, final point and symmetry. Draw the trajectory on the blackboard, predict its shape to be a parabola
# 3 15 min	Verify the hypothesis by mathematical formulation	The teacher guides students to carry out mathematical deduction using the kinematic equations for horizontal and vertical motion Determine the displacement of the projectile along the x-axis and y-axis to deduce its trajectory	Group activity, guided by teacher	Blackboard, Worksheet 1	– Plot the trajectory on the blackboard – Guide students to choose appropriate coordinates and draw the vectors – Review kinematic equations of motion with constant acceleration, if necessary

No./Time	Stage	Learning activities	Interaction form	Materials/ Resources	Methodological comments
# 4 10 min	Verify mathematical deduction with experiment	The teacher uses software to analyze the video in order to curve fit the trajectory of the projectile. Conclude that it is a parabola	Lecture	Laptop, video analysis software, projector	– Use the videos from the beginning of the lesson to analyze the shape of the trajectory and determine the curve fit
# 5 15 min	Determine range and maximum range	The teacher asks students to find the intersection of the parabola and the x-axis, the vertex of the parabola, and its critical point and analyze factors affecting the vertex and the range. Thus, students can draw conclusions about the conditions needed to achieve maximum range	Teacher presents; students work individually and discuss under the guidance of the teacher	Blackboard, Worksheet 2	– Present that the trajectory of a projectile is a parabola corresponding to the kinematic equation derived earlier in the lesson – Give this hint to find the range: the final horizontal displacement is double that of vertex (because of the symmetry of the parabola)

No./Time	Stage	Learning activities	Interaction form	Materials/ Resources	Methodological comments
# 6 15 min	Verify the formula to find the maximum range	The teacher carries out the verification experiment using the water stream experiment kit, with which one can adjust the initial angle and fix the initial velocity	Teacher presents and carries out the experiment Students observe	Water stream experiment kit	• Introduce the experiment Prepare a table with the angles 30, 40, 45, 50, and 60 degrees and the range for each angle • Using the table, the maximum range can be derived (45° angle) and the relationship confirmed
# 7 5 min	Confirm the experimental results and discuss further	The teacher summarizes on the blackboard the kinematic equations of motion, how to determine the range, and the conditions that produce maximum range Then the teacher discusses how to determine velocity, the symmetry of velocity at the same height, and when the object is thrown at a height compared to from the ground	Teacher gives a lecture Students listen and take notes	Blackboard, laptop, projector, presentation	While presenting, ask questions: What is the velocity of the projectile at height h when it is traveling upward? When traveling downward? How are the velocities along Ox and Oy and the angle at the vertex different from those at the final point?

No./Time	Stage	Learning activities	Interaction form	Materials/ Resources	Methodological comments
# 8 10 min	Practical application	Students are asked to determine the initial angle that Olympic champions use in shot put and javelin throw	Work in groups	Videos and analysis software	Teacher lets students work in groups by themselves. If they cannot finish by the end of the class, let them finish at home

Worksheet 1: Derive the equation of projectile motion

Name: ——————————————

Group: ——————————————

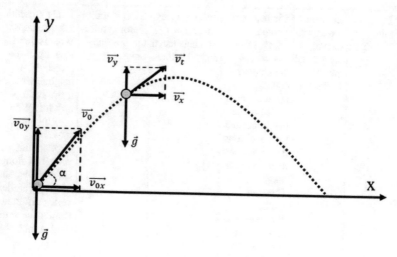

An object is thrown with the initial velocity $\vec{v_0}$ and forms angle α with the horizonal axis. Choose a coordinate system Oxy: origin O is the initial position, Ox is the horizontal direction of motion, and Oy is the vertical direction.

Analyzing the initial velocity along Ox and Oy gives

$$\vec{v_0} = \vec{v_{0x}} + \vec{v_{0y}}, \tag{11.1}$$

with the magnitude

$$v_{0x} = v_0 \cdot cos\alpha; \; v_{0y} = v_0 \cdot sin\alpha; \; v_0^2 = v_{0x}^2 + v_{0y}^2 \tag{11.2}$$

The motion of the projectile along Ox $\{v_{0x} \neq 0; a_x = 0; x_0 = 0\}$ is motion at a constant speed.

Write the kinematic equation of motion that describes the motion of the projectile along Ox:

$v_x=$ _____ (I)

$x =$ _____ (II)

The motion of the projectile along Oy $\{v_{0y} \neq 0; a_y = -g; y_0 = 0\}$ is motion at a constant rate of acceleration.

Write down the kinematic equation of motion that describes the motion of the projectile along Oy:

$v_y=$ _____ (III)

$y =$ _____ (IV)

The velocity vector of a projectile at any given time is the sum of the velocity vectors along Ox and Oy: $\vec{v_t} = \vec{v_x} + \vec{v_y}$ and $v_t^2 = v_x^2 + v_y^2$. The angle is determined by $tan\alpha = \frac{v_y}{v_x}$.

From expressions (I) and (III), derive the expressions for the velocity of the projectile and the formula to calculate angle α:

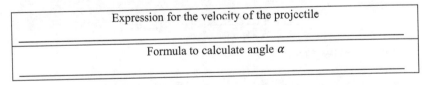

Expression for the velocity of the projectile
Formula to calculate angle α

From expressions (II) and (IV), derive the expressions for the relationship between x and y and the trajectory of the projectile:

Expression of the relationship of x and y
Trajectory of the projectile

Worksheet 2: Determine the maximum range of a projectile

Name: ———————————

Group: ———————————

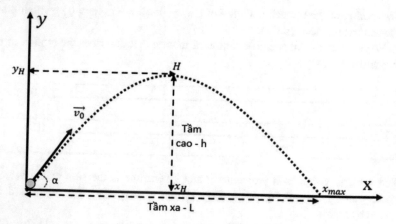

The trajectory of a projectile is a parabola corresponding to the equation of motion:

$$y = -\left(\frac{g}{2v_0^2 (cos\alpha)^2}\right)x^2 + (tan\alpha)x,$$

when it touches the ground:

$$y = 0 \rightarrow -\left(\frac{g}{2v_0^2 (cos\alpha)^2}\right)x^2 + (tan\alpha)x = 0 \qquad (11.3V)$$

Solving equation (V) gives the range of the projectile.

> Expression to calculate the range of the projectile:
>
> _____
>
> _____
>
> _____

> The range of the projectile depends on the following factors:
>
> _____
>
> _____

Based on the range of the projectile, determine the conditions for maximum range:

> Maximum range is reached when:
>
> _____
>
> _____

Suggestion: Water stream experiment

Take a transparent 1.5 L bottle, drill a hole in the bottom, and insert a long, thin, flexible tube. Drill a hole in the cap of the bottle and insert another tube until it nearly reaches the bottom, so that when water leaves the first tube, the initial velocity will not change.

Place the tip of the first tube in the center of a quarter circle drawn on the board with angle markings (0–90 °). Attach a ruler to the board horizontally to determine the range of the water.

Place a container on the ground to collect the water.

Table

Angle (degree)	30	40	45	50	60
Range (cm)					

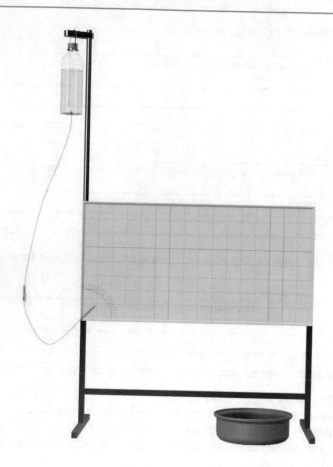

Lesson Plan: The Dependence of Resistance on Temperature

12

Tran Ngoc Chat

Lesson Title: The dependence of resistance on temperature
Abstract: In this lesson, students will learn the relationship between resistance and the temperature of a conductor. The main part of the lesson will investigate metal conductors with a first-order functional relation. Through measurements ob-tained from experiments, students will construct a function representing the relationship between resistance and temperature. Through this connection, stu-dents will be able to solve various practical problems.

Type of school / Grade	High School / Grade 11
Prerequisites	Mathematics: Characteristics of first-order functions Physics: – Ohm's law – Knowledge of how to use digital multimeters
Number of periods	2

T. N. Chat (✉)
Faculty of Physics, Hanoi National University of Education, Hanoi, Vietnam
e-mail: chattn@hnue.edu.vn

© The Author(s), under exclusive license to Springer Fachmedien Wiesbaden
GmbH, part of Springer Nature 2022
F. Dilling and S. F. Kraus (eds.), *Comparison of Mathematics and Physics
Education II*, MINTUS – Beiträge zur mathematisch-naturwissenschaftlichen
Bildung, https://doi.org/10.1007/978-3-658-36415-1_12

Objectives	Mathematics Students can: – Draw a first-order graph representing the relation-ship of volt-age and current, according to Ohm's law – Detect that the gradient of the U-I graph is the resistance – Detect that steeper the U-I graph, the greater the U or I value – Identify the characteristics of the function from the table of measured values R and T – Propose a graph representation plan to find the function for the relationship of R and T – Identify the first-order function from the points on the graph corresponding to the measured values R and T – Generalize the function that shows the relationship between R and T. Physics Students can: – Apply Ohm's law to predict current values in a circuit – Propose a hypothesis about the relationship between R and T to explain the contradiction between predicted and measured results – Propose a reasonable plan to study the relationship between R and T – Assemble simple electrical components according to electrical circuit diagrams – Adjust the circuit's parameters to measure the appropriate values – Apply conclusions regarding the relationship between R and T to solve practical problems

12.1 Methodical Commentary

In the junior high school physics program, students have learned Ohm's law: voltage U is proportional to current I with a scale factor of $R = U/I$. R is a constant value called resistance. For Ohm's law, the simple case is one in which electrical components have a constant value for R. Therefore, students will have the misconception that the voltage remains proportional to the current streaming through the filament bulb. However, in this lesson, when doing experiments with filament light bulbs, students will see that I is not directly proportional to U; instead, the greater U and I are, the greater R is. R does not remain constant.

Students also know that, with a greater current, the conductor has a higher temperature. Based on this, students will be able to hypothesize that the conductor's resistance depends on temperature. However, they will not yet know what this relationship looks like. This contradiction will motivate students to learn.

To explore this relationship, students will need to propose a plan, which involves designing and conducting an experiment to gather data. From the data table, it will be possible to determine the hidden mathematical rules that can be represented by functions. Students will conduct an experiment to investigate the relationship between R and T. From the table of data obtained from the experiment, students will draw an R-T graph. Using the mathematical knowledge that they have learned, they will find the functions representing the relationship of R and T.

After finding this relationship, students will learn the meaning of the coefficient of thermal resistance and apply the laws to solve problems associated with practical and technical applications.

12.2   Lesson Plan

Nr./ Time	Stage	Learning activities	Interaction form	Materials/ Resources	Methodological comments
#1 5 min	Ensure background knowledge	The teacher assigns students the problems: What are some characteristics of the thermal effect of electric current? State the content of Ohm's law. Write the expression of the law	Teacher present/ Whole class work		Make sure students have mastered the simple case of Ohm's law. Based on this, problems will be introduced in the following learning stages
#2 15 min	Detecting the problem	The teacher divides the class into four groups Students conduct Experiment 1. The teacher reads the experimental results of the first pair of U and I values. Students predict the next U and I values Students conduct experiments to measure pairs of U and I values. They compare the measurement results and predicted values. Ask students to comment Students explain the difference between prediction and measurement The teacher formally states the problem	Group work	Worksheet 1 Experiment 1	Students have learned Ohm's law, U is proportional to I with a scale factor of $R = U/I$ and R is called resistance. Therefore, students will predict experimental data according to the proportionality of U and I and fill in the table. The graph representing U versus I will be a straight line with a constant slope However, when compared with the experimental results, the data show a significant deviation. When represented graphically, the experimental curve lies higher than the predicted line. The greater the value of U or I, the greater the discrepancy between experiment and prediction; the slope of the experimental graph changes continuously

Nr./ Time	Stage	Learning activities	Interaction form	Materials/ Resources	Methodological comments
					Because students already know that the greater the current, the higher the temperature of the conductor, they can hypothesize that the resistance of a conductor depends on temperature So, the problem is: What is the relationship between conductor resistance and temperature?
#3 15 min	Proposing a plan to solve the problem	The teacher asks prompting questions so that students can suggest solutions Students propose solutions: get experimental data, represent that data in the form of graphs, and find the relationship of R and T The teacher introduces the experiment set to be given to students Students learn about experimental equipment. The teacher demonstrates how to set it up and conduct experiments	Teacher present/ Whole class work		The teacher suggests to the students that when they need to find the specific relationship of two quantities, they need a table of data. Through the graphical representation of the data, it is possible to find a mathematical relationship (function) between the two quantities To create that data table, it is necessary to conduct experiments to measure pairs of quantities R and T Therefore, it is necessary to first design an experimental plan to investigate the relationship between R and T Students will see that R and T have to be measured. Therefore, it is possible to propose an experiment involving a resistance R. To measure resistance requires a conductor, a circuit with a power source, voltmeter, ammeter, and connecting wires. To measure the temperature of the resistance R, one dips the resistance R into a cup of water and measures the temperature of the water. To change the temperature of the water, one can add ice or hot water. A kettle for hot water and a container of ice are also necessary pieces of equipment to prepare

Nr./ Time	Stage	Learning activities	Interaction form	Materials/ Resources	Methodological comments
#4 25 min	Implement the solution	Students conduct experiments and collect experimental data Process the experimental results by plotting R against T. From the graph, draw conclusions about the relationship between R and T	Group work	Experiment 2	After students collect experimental data and represent the data on a graph, it is easy to realize the linear relationship between R and T: $R_T = a \cdot T + R_o$ where a is a constant, as well as the gradient of the curve $R\text{-}T$
#5 5 min	Conclusions	The teacher guides students to recognize the consistency of research results and those in the textbook The teacher determines the conclusion of the lesson and defines the quantity Students learn the coefficient of thermal resistance Students state and take notes on important conclusions of the lesson	Teacher present/ Whole class work		The relationship that students found is similar to the conclusion in the textbook: $R_T = R_o \cdot (1 + \alpha \cdot (T - T_o))$ where α is called the temperature coefficient of resistance
#6 10 min	Practice	The teacher corrects exercise #1 on Worksheet 3 Students practice solving some basic problems on Worksheet 3	Teacher present/ Group work	Worksheet 3	The teacher demonstrates a sample solution to an illustrated problem to help students familiarize themselves with new concepts and the type of problem
#7 15 min	Extention and application	The teacher asks students to explain the reason why the metal resistance increases when the temperature increases, based on physical theories of atomic molecules The teacher introduces materials that have relationships between R and T that follow other rules	Teacher present/ Whole class work	Experiment 3	The teacher broadens the students' knowledge by asking them to use theories of atomic molecules and the conductive properties of metals to explain the increase in electrical resistance with an increase in temperature. Thereby, it can be seen that the content of physics knowledge is uniform

Nr./ Time	Stage	Learning activities	Interaction form	Materials/ Resources	Methodological comments
		The teacher asks students to design an application and observe an illustrated experiment Students propose an application plan, predict experimental results, observe experimental results, and compare with their predictions			Introducing other materials with very great or negative temperature coefficients of thermal resistance helps students see the achievements of physics research and shows the practical application of the fact that the electrical resistance depends on temperature (e.g., temperature measuring devices, automatic disconnection of electrical equipment when overheated, etc.)

Worksheet 1

Investigate the relationship between the voltage U and the amperage I of the lamp filament according to the following circuit:

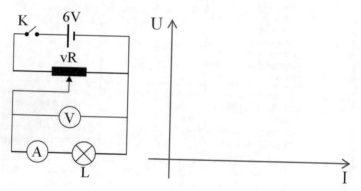

Task 1: Conduct the experiment once. Record the values of the first measurements U_1 and I_1. Based on Ohm's law, predict the value of amperage I at the next measurements. Plot the predictions on the coordinate plane (Table 12.1).

Table 12.1 Prediction of experimental results

No	1	2	3	4	5	6	7	8	9	10
U (V)	0.1	0.2	0.4	0.8	1	2	3	4	5	6
I (mA)										

Table 12.2 Measurement results from the experiment

No	1	2	3	4	5	6	7	8	9	10
U (V)	0.1	0.2	0.4	0.8	1	2	3	4	5	6
I (mA)										

Task 2: Conduct the experiment to measure the amperage values of the following measurements. Record them in the table of values. Plot the measured values on the same coordinate plane as the prediction graph (Table 12.2).

Worksheet 2
Experiment 2: Investigate the relationship between resistance and temperature.

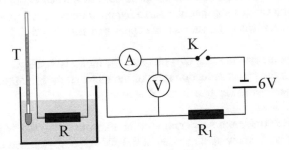

Task 1: Change the temperature of the water. Record the values of voltage U, current I, and temperature T in the data table. Calculate the corresponding resistance values R.

No	1	2	3	4	5	6	7	8	9	10
T (°C)										
U (V)										
I (mA)										
R (Ω)										

Task 2: Draw a graph showing the relationship between R and T.

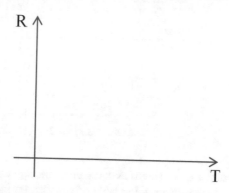

Task 3: Describe the gradient of the line graph. Derive a function that represents the relationship between R and T.

Worksheet 3

Problem 1: A copper coil in a DC motor at 25 °C has a resistance of $R_o = 10 \, \Omega$. When the motor is running stably, it has a temperature of 50 °C. Knowing that the temperature coefficient of resistance of copper wire is $\alpha = 4,5 \cdot 10^{-3} \left(K^{-1} \right)$,

a) Determine the resistance of the coil when the motor is running.
b) Determine the power reduction ratio of the motor at the initial time compared to the time when the engine is operating stably.

Problem 2: Determine the temperature of the filament of a 6 V-6 W incandescent bulb when it is lit, knowing that, when a 100 mV voltage is applied to the bulb at a temperature of 0 °C then the current through the bulb is 210 mA. The Wolfram filament has a temperature coefficient of resistance of $\alpha = 4,5 \cdot 10^{-3} \left(K^{-1} \right)$.

  Problem 3: A induction cooker creates a constant electromotive force on the bottom of a pan, regardless of what is being cooked. Consider two cases: In the first case, when the pan is used to cook soup, the bottom reaches a temperature of 100 °C; in the second case, when it is used for frying, the bottom reaches a temperature of approximately 220 °C.

a) In which case in which case is the pan provided with greater heat power? Why?
b) Determine the ratio of the difference in heat power supply in the two cases.

Experiment 3. Illustrating the application of protecting electrical equipment from overheating.

The experimental setup is a circuit diagram where R_T is a thermistor with a high temperature coefficient of resistance, which increases when the temperature increases.

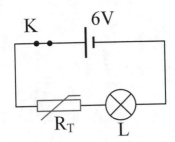

When switch K is closed, the light L is on.

When we bring the candle flame close to the thermistor, we see that the light fades and turns off.

When we take the candle away from the thermistor, we see that after about one minute, the lamp L slowly lights up again.

Explanation: The thermistor in the experiment is a PTC (positive temperature coefficient) type. Unlike metal, this thermistor is made of a special material, which has the property that when the temperature increases, its temperature coefficient of resistance increases, so the resistance increases significantly (compared to the resistor made of metal). Therefore, when the candle flame gets close to the thermistor, the temperature coefficient of resistance increases rapidly, causing the current in the circuit to drop rapidly and the light to become weaker and then go out. The process is reversed when the candle flame is moved away from the thermistor.

Lesson Plan: Simple Harmonic Oscillation

13

Nguyen Van Bien

Lesson Title: Simple harmonic oscillation
Abstract: This lesson covers harmonic oscillation, a concept from physics used to describe periodic processes such as the oscillation of a pendulum. Periodic functions from mathematics play a significant role in this lesson.

Type of school / Grade	High School / Grade 12
Prerequisites	– Knowledge of trigonometric functions – Basic understanding of kinematics
Number of periods	1
Objectives	Mathematics: – Students can apply their knowledge of periodic functions to physical concepts (harmonic oscillation) Physics – Students know the principle of a harmonic oscillator and related concepts – Students describe periodic processes as harmonic oscillators and can perform simple calculations using this knowledge

N. V. Bien (✉)
Faculty of Physics , Hanoi National University of Education, Hanoi, Vietnam
e-mail: biennv@hnue.edu.vn

© The Author(s), under exclusive license to Springer Fachmedien Wiesbaden GmbH, part of Springer Nature 2022
F. Dilling and S. F. Kraus (eds.), *Comparison of Mathematics and Physics Education II*, MINTUS – Beiträge zur mathematisch-naturwissenschaftlichen Bildung, https://doi.org/10.1007/978-3-658-36415-1_13

13.1   Methodical Commentary

The lesson presented here is an introduction and a first stage of practicing the principles of harmonic oscillation. The lesson is suitable for both mathematics and physics classes, with a shifting emphasis depending on the subject. For mathematics, the focus is on periodic functions and their application in scientific theories. For physics, the focus is on the idea behind the principle and the phenomenon.

13.2   Lesson Plan

Lesson Plan

No./ Time	Stage	Learning activities	Interaction form	Materials/ Resources	Methodological comments
# 1 15 min	Define the problem	Teacher gives examples of mechanical oscillations with videos, and then assigns students tasks: + What are the common characteristics of vibrating objects? Give some examples + Discuss the periodic oscillation of the pendulum in a timepiece or select a periodic oscillation in the examples given by the students, and design an experiment to verify the prediction Students perform the tasks and report the results Teacher summarizes: + Mechanical oscillation: The back-and-forth movement of an object around a particular position (equilibrium position) is called "oscillation." + Periodic oscillation: If, after equal intervals of time, the object returns to its original state, the oscillation of the object is said to be "periodic."	Teacher asks/individual answers	https://www.youtube. com/watch?v= 6O0mjlpag6M https://www.youtube. com/watch?v= p9VZV6HTLao	These activities motivate students to recognize the features of mechanical oscillations and learn the concepts of amplitude, period, frequency, angular frequency, and phase difference to describe harmonic oscillation Students are expected to be able to give examples of mechanical oscillations and to design and execute an experiment to explain periodic oscillation Expected results from students: + Examples of mechanical oscillation should be objects that move back and forth around the equilibrium position (the position where the object is at rest). Examples can include a hammock, a clock pendulum, a boat bobbing, leaves swaying in the wind, guitar strings vibrating, or drums vibrating + Discuss the oscillation of the clock pendulum:

(Fortsetzung)

No./ Time	Stage	Learning activities	Interaction form	Materials/ Resources	Methodological comments
		+ Period: The amount of time the object takes to make one complete oscillation is called the "period" (or the shortest time the object takes to return to the same position in the same direction). Symbol: T; cycle unit(s) + Frequency: Number of oscillations made in 1 s + Amplitude: The position of the farthest displacement from the equilibrium position during the motion of the object **The problematic situation:** Due to the different ways an object might oscillate, we classify oscillations into types, including free oscillation, periodic oscillation, and harmonic oscillation *What are the features of harmonic oscillations?*			Prediction: After equal intervals of time, the pendulum returns to its original state + Propose an experimental plan to verify the prediction: Measure the intervals of time as the pendulum passes through any landmark position and plot an angle-time graph Inline + Students report and confirm the prediction is correct. Teacher assigns: What is the term for the shortest time it takes an object to return to its original position?
# 2 15 min	Solve the problem / Develop new knowledge	Teacher assigns tasks: + Discuss the graph of pendulum oscillation in Activity 1 + Write the equation for the oscillation of the pendulum from the experiment in Activity 1 Teacher gives the definition of harmonic oscillation:			

(Fortsetzung)

No./Time	Stage	Learning activities	Interaction form	Materials/Resources	Methodological comments
		+ Harmonic oscillation is an oscillation in which the change in displacement over time is a sine or cosine function $x = A.\cos(\omega t + \varphi)$, where A: amplitude of oscillation; ω: angular frequency (rad/s); φ: initial phase (rad) Teacher asks students to describe harmonic oscillations in some examples using learned concepts Teacher assigns tasks: Build the equation for velocity and acceleration of the harmonic oscillating object, knowing that velocity is the derivative of displacement with respect to time, and acceleration is the derivative of velocity with respect to time Students perform the tasks Teacher summarizes: + Apply the equation of velocity as the derivative of displacement to build the equation $v = -A\omega\sin(\omega t + \varphi)$ Teacher makes comments: The velocity of harmonic oscillation varies with the same frequency, faster by $\pi/2$ in phase with respect to the displacement			

No./Time	Stage	Learning activities	Interaction form	Materials/Resources	Methodological comments
		The minimum speed of harmonic oscillation is 0, at the boundary position. The maximum speed of the harmonic oscillation is $A\omega$, at the equilibrium position. + Applying the equation of acceleration, which is the derivative of the velocity, build the equation $a = -A\omega^2 cos(\omega t + \varphi) = -\omega^2 x$ Make comments: The acceleration of harmonic oscillation varies with the same frequency and opposite phase with respect to the displacement. The minimum magnitude of acceleration of the harmonic oscillation is 0, at the equilibrium position. The maximum magnitude of acceleration of harmonic oscillation is $A\omega^2$, at the boundary position			
# 3 7 min	Practice the knowledge	Students solve relevant problems by applying equations of harmonic oscillation under teacher instruction	Individual working	Task 1	
# 4 8 min	Elaboration	Students apply the concepts of amplitude, period, frequency, angular frequency, and phase difference to describe harmonic oscillation with guidance from teacher. Students use learned knowledge to recognize the mechanical oscillations in life and solve practical problems with guidance from teacher	Individual working	Task 2	

Group tasks:
Task 1

Problems:

Problem 1: Find the relationship between harmonic oscillation and uniform circular motion.

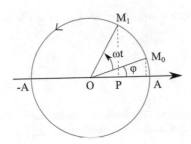

Problem 2: The equation for the harmonic oscillation is $x = 2cos\left(5t - \frac{\pi}{6}\right)$ cm.

a) Show the amplitude, initial phase, and phase at time t.
b) Write equations for velocity and acceleration.

Solutions:

Problem 1:
An object P oscillating harmonically on a line segment can be considered the projection of a corresponding particle M in uniform circular motion on the diameter of that line segment, as the equation for point P is $x = A \cdot \cos{(\omega t + \psi)}$

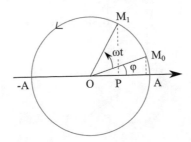

Problem 2:

a) Amplitude: $A = 2$ cm

Initial phase: $\varphi_0 = \frac{\pi}{6}$ rad

Phase at time t: $\varphi = 5t - \frac{\pi}{6}$ rad

b) Equation of velocity: $v = -A\omega sin(\omega t + \varphi) = -10\ sin\left(5t - \frac{\pi}{6}\right)$ cm/s

Equation of acceleration: $a = -A\omega^2 cos(\omega t + \varphi) = -\omega^2 x = -50cos\left(5t - \frac{\pi}{6}\right)$ cm/s^2

Task 2

Problems:

Problem 1: Give examples of mechanical oscillations in real life, and state which oscillations are periodic, non-periodic, and harmonic.

Problem 2: The piston of an internal combustion engine oscillates in a straight line 16 cm long and causes the engine crankshaft to rotate at a constant speed of 1200 rpm (Figure below).

a) Write the equation of displacement of the piston.
b) What is the maximum speed of the piston, and at what position?
c) What is the maximum acceleration of the piston, and at what position?

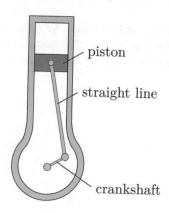

Solutions:

Problem 1:
Real-life examples of motor oscillation include a swing, a buoy bobbing on the water, a swinging bridge, a pendulum clock, and a piston in an engine.

Harmonic oscillation in the engines, the oscillation above is a pendulum clock, the piston oscillates.

Problem 2:
If the distance from the piston to the projection of the crankshaft onto the cylinder shaft is approximately equal to the length of the boundary, which is constant, the piston oscillates approximately as the projection of the crankshaft onto the cylinder shaft. That oscillation has a frequency f equal to the crankshaft rotational frequency: $f = \frac{1200}{60} = 20$ Hz.

The amplitude A is $A = \frac{0,16}{2} = 0,08$ m.

a) Choose the time origin $t = 0$ at the maximum displacement $x = A$, which is when the piston is at its highest position and the positive direction of the x-axis is upward; the equation of vibration of the piston is $x = A\cos\omega t = A\cos 2\pi f t = 0,08\cos 40\pi t \,(\text{m})$.

b) The velocity has a maximum magnitude of $v_{max} = A\omega = 10$ m/s at position $x = 0$.

c) Acceleration has a maximum magnitude of $a_{max} = A = 2 = 1262$ m/s^2 at the boundary position $x = \pm\,0,08$ m.

Note: In a motor, the acceleration of the piston and the actuator connected to it is immense. Thus, the machine parts must have enough durability to withstand the intense force.

Comparison: Vectors in Mathematics and Physics Education

14

Frederik Dilling

14.1 Introduction

The concept of vectors was initially developed to describe geometrical and physical phenomena (cf. Filler, 2011). Therefore, a so-called Euclidean vector was regarded as a quantity that had both a direction and a magnitude. In Euclidean geometry, vectors were defined in the nineteenth century as an equivalence class of ordered pairs of points. Two pairs, (A, B) and (C, D), are equivalent if points A, B, D, and C in that order form a parallelogram. The notation $\overrightarrow{AB}$ is used for the equivalence class. In physics, vectors describe directed quantities, those that have a direction and a magnitude, e.g., velocities and forces. Thus, the concept of vectors has developed in the interplay of mathematics and physics:

> "The vector calculation was developed in a long historical process, mainly due to the need for a geometric calculation and the requirements of physics." (Filler, 2011, p. 85, translated)

With the formalization of the foundations of mathematics in the twentieth century, the concept of vectors was described in the language of set theory. Since that time, vectors in mathematics as a scientific discipline have been defined by vector space axioms.

Mathematics teaching in schools uses vectors in analytical geometry and linear algebra. Vectors are usually described in a similar way to the definition of

F. Dilling (✉)
Mathematics Education, University of Siegen, Siegen, Germany
e-mail: dilling@mathematik.uni-siegen.de

© The Author(s), under exclusive license to Springer Fachmedien Wiesbaden GmbH, part of Springer Nature 2022
F. Dilling and S. F. Kraus (eds.), *Comparison of Mathematics and Physics Education II*, MINTUS – Beiträge zur mathematisch-naturwissenschaftlichen Bildung, https://doi.org/10.1007/978-3-658-36415-1_14

the Euclidean vector: an equivalence class of arrows of equal length and direction (see Fig. 14.1). Alternatively, vectors are arithmetically defined as n-tuples of real numbers. The formalistic version as vector space axioms is not used in teaching. Various applications of vectors are discussed in physics classes. Especially in mechanics, students encounter directed quantities (see Fig. 14.2).

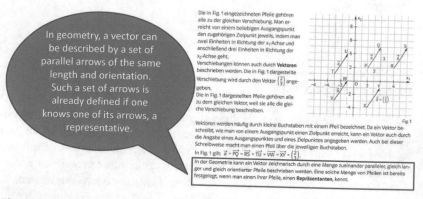

Fig. 14.1 Introduction of vectors in a German mathematics textbook in a geometrical context (Brandt et al., 2014, p. 116)

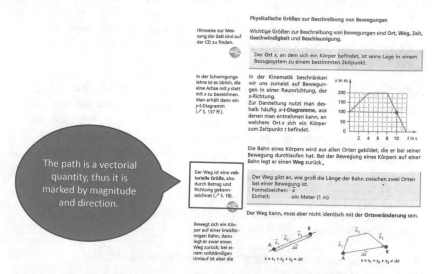

Fig. 14.2 Introduction of vectors in a German physics textbook in the context of translations (Hoche et al., 2003, p. 58)

Since vectors are used both in mathematics and in physics, educational research in both subjects deals with this topic. In this chapter, some important findings of mathematics education research and physics education research on vectors will be explained and brought together. The text is substantially based on Dilling (2019a).

14.2   Mathematics Education Research

14.2.1   Approaches to the Concept of Vectors

In linear algebra and analytical geometry, mathematics education research has intensively investigated different approaches to the concept of vectors. In particular, three approaches can be distinguished (cf. e.g., Filler & Todorova, 2012; see Fig. 14.3): vectors can be introduced as arrow classes, as n-tuples, or by vector space axioms. This section will present these three approaches in more detail.

The approach via *arrow classes* is performed by defining a vector as a class of arrows of equal length, direction, and orientation. This geometric approach is of great importance for students' understanding of vectors because the essential properties of the concept can be related to geometrical objects (cf. Dilling, 2021). However, the approach via arrow classes is also mentioned critically in the educational discussion (see, e.g., Malle, 2005a). Students naturally think in points and arrows, whereas arrow classes constitute a complex concept that is difficult to apply. Furthermore, many vectors in physics are not arrow classes (see Sect. 3.2). Additional difficulties arise in the distinction between class and representative and in the complex introduction of arithmetic operations within this concept.

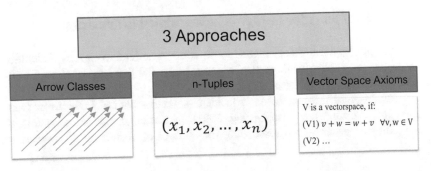

Fig. 14.3 Approaches to the concept of vectors

Therefore, the concept of arrow classes tends not to be consistently used in class beyond the introductory stage.

The approach via n-tuples uses vectors as n-tuples of real numbers. These can subsequently be used to describe geometric situations. The vectors can be interpreted as points and arrows, and they do not have to be understood as arrow classes. That facilitates the introduction of arithmetic operations for vectors. The access via n-tuples often leads to the difficulty that this arithmetic concept fades away in a student's long-term development and is replaced by geometric perceptions. Therefore, the transition between an arithmetic and a geometric concept of vectors should be practiced explicitly in class (cf. Bender, 1994).

Another approach to introducing vectors is via vector space axioms. This approach relates to a general axiomatic structure of mathematics teaching. Since the new math movement (see, e.g., Kilpatrick, 2012), such a structure is no longer used in schools but is reserved for courses in linear algebra at universities. Although the approach is technically elegant, its complexity makes it unsuitable for the first contact in mathematics lessons at school.

Dilling (2019b) has investigated the introduction of the concept of vectors in German textbooks for mathematics and physics. In all four textbooks considered in that empirical study, the concept of vectors was introduced via arrow classes. Only one of the mathematics textbooks additionally defined vectors as n-tuples. Besides the introduction of the concept, it became apparent that, in the physics textbooks, no n-tuples were used at any point. Instead, drawn arrows were used to describe physical concepts and to determine concrete values. However, in the mathematics textbooks, numerous n-tuples were used to describe various situations. That shows that the concept of vectors is introduced and used differently in each of the two subjects.

14.2.2 Problems for Students in Working with Vectors

The complex and multifaceted concept of vectors can lead to various difficulties in dealing with these mathematical objects. Malle (2005b) has listed such challenges, which will be discussed in this section.

A widespread misconception is to identify vectors with a single arrow (*vector = single arrow*) (see also Dilling, 2021). In particular, this interpretation of a vector can occur when introducing the concept of vectors as arrow classes.

Instead of considering a single arrow as a representative of the arrow class, the single arrow itself is mistakenly regarded as the vector.

A further problem in dealing with vectors is the strong presence of geometric ideas. *Vectors are interpreted geometrically* by many students; an arithmetic concept as a pair of numbers, which is necessary for various applications, is rarely accepted by students. Furthermore, many students believe that arrows cannot be indicated by one pair of numbers. Instead, they think the start and end points of an arrow must always be indicated.

Students also have problems with the so-called *position vector*. That is a concept specially developed for teaching to avoid a distinction between the addition of two vectors and the addition of a vector and a point. Therefore, points are defined as position vectors. That often causes students to interpret vectors as position arrows with the starting point at the origin. It also leads to difficulties in understanding the *zero vector*. Many students think that the zero vector cannot be an arrow but only a point.

Various other problems can be observed in the classroom when dealing with vectors. For example, many students cannot correctly set up formulas with vectors; they understand the task but fail to convert the information into a formula. Students often perform vector calculations correctly, but they do so without understanding and by using a method that is unrelated to the originally learned concept.

14.3   Physics Education Research

14.3.1 Arrows (as Vectors) in Physics Class

Physics education research on vectors often refers to the use of arrows in physics (see, e.g., Boczianowski, 2012). Students already have significant experience with arrows in everyday life: for example, the hands of a clock or the arrows on street signs. Textbooks for other subjects also use representations with arrows: for example, annotated maps in geography lessons or flow charts for the description of processes (see Fig. 14.4). However, the arrows used in physics classes, including the vector arrow, follow different rules than such arrows. That can lead to difficulties in teaching if the ideas and rules are transferred without appropriate adaptation.

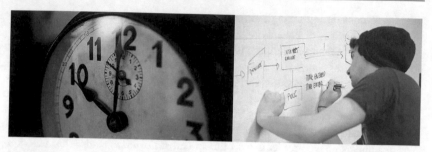

Fig. 14.4 Arrows as the hands of a clock and within a flow chart

The use of arrows is a sustainable concept in physics lessons. Arrows can be used to model problems associated with directed quantities (e.g., velocity and acceleration) or periodic quantities (e.g., the amplitude of electromagnetic waves) without the need for an arithmetic description. In this context, the magnitude of the directed quantity is defined by the length of the arrow. That makes the process clear and concise. Limitations arise when dealing with variables because only a certain number of arrows can be represented at the same time.

Directed quantities are a central topic in physics education, especially in mechanics (see Sect. 1.3.2). However, students often have difficulty distinguishing between a directed and an undirected quantity. That can be explained by the frequent limitation to a one-dimensional case, where the distinction between the vector, the vector component, and the magnitude is not obvious. For example, "speed" is a scalar quantity with the dimension length per unit of time, whereas "velocity" is a vectorial quantity or a vector component with the same dimension.[1] Therefore, changes in directed and undirected quantities should be specifically addressed, and the correct terms should be used consistently to prevent misconceptions (cf. Vu et al., 2020). Directed quantities should be described at an early stage in multidimensional situations by arrows and should be used consistently in later units of the course.

Various problems can be identified when students deal with illustrations with vector arrows (cf. Kraus, 2016). A frequent problem is the selection of aspects in

---

[1] Such a distinction is not made in the German language. The word *Geschwindigkeit* describes the vectorial as well as the scalar quantity.

representations. Single arrows are not shown to emphasize other ones. For example, opposing forces are often not considered in plots, which can lead to misconceptions. Another problem is the so-called polysemy. If different directional quantities are similarly shown in one representation, this can lead to problems in distinguishing and interpreting the single objects. Furthermore, many students interpret arrows as changes over time, leading to incorrect interpretations in the case of vectors. Thus, a force diagram may well represent a static (i.e., time-invariant) situation.

14.3.2 Application of Vectors in Mechanics

In physics classes, students usually encounter vectors for the first time in mechanics (cf. Boczianowski, 2012). The different directional quantities are represented as arrows. Therefore, different properties are bound to these arrows.

The introduction of the concept of directional quantities usually occurs in the context of translations. In a translation, the shaft of the arrow represents the distance moved. Another type of vector is velocity. In this case, the arrow is oriented in the direction of the movement, and the length represents the current speed. That can lead to difficulties because it causes a blending of coordinate space and speed space. Even more properties are attributed to the arrow when using the concept of force. The arrow of force is oriented in the direction of the force, the length represents the amount of the force, and the base of the arrow is the point of impact of the force (see Fig. 14.5). Thus, it is not a vector, i.e., a class of arrows in the mathematical sense. This definition of force also makes it difficult to handle more complex applications with different points of impact. It would therefore make sense to avoid defining the base of the arrow as the point of impact.

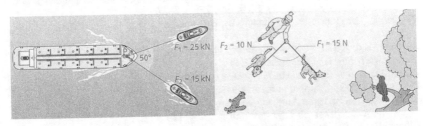

Fig. 14.5 Decomposition of forces in a German physics textbook (Blüggel et al., 2020, p. 117)

14.4 Conclusion and Outlook

This article has demonstrated that mathematics education research and physics education research investigate different facets of the concept of vectors. It is mathematics that deals with the fundamentals of the concept. A distinction is made between the approaches as arrow classes, as n-tuples, and via vector space axioms. Furthermore, various difficulties in dealing with vectors have been described. Physics is primarily concerned with the application of the concept of vectors to describe physical phenomena. It turns out that different directional quantities are transferred to vectors with entirely different properties. The representation of vectors by arrows is especially important in physics lessons. Therefore, it is crucial to consider the previous experiences of students with arrows.

The approaches developed in mathematics education research and physics education research can also be usefully applied in the other subject. Students often encounter vectors for the first time in physics classes in the context of translations. Therefore, basic knowledge about various approaches to the concept of vectors is helpful for physics teachers. In mathematics classes, physical examples and other applications are often discussed in addition to the concept of vectors itself. For this reason, mathematics teachers should know the correct handling of these applications, e.g., the distinction between speed as a scalar and velocity as a vectorial quantity, as described in Sect. 1.3.2. A well-planned mathematics and physics education that considers the specifics of the other subject can prevent the development of isolated concepts.

References

Bender, P. (1994). Probleme mathematischer Begriffsbildung diskutiert am Beispiel der Vektor-Addition. *Mathematica Didactica, 17*(1), 3–27.
Blüggel, L., Hegemann, A., Feldmann, C., & Kohl, R. (2020). *Impulse Physik* (p. 7–10). Klett.
Boczianowski, F. (2012). Pfeile als themenübergreifendes Symbolsystem im Physikunterricht. *PdN Physik in Der Schule, 61*(4), 5–10.
Brandt, D., et al. (2014). *Lambacher Schweizer Mathematik. Einführungsphase.* Klett.
Dilling, F. (2019a). Vektoren in der mathematik- und physikdidaktischen Forschung. Stoffdidaktischverbindende Ansätze zwischen linearer Algebra und klassischer Mechanik. *Phydid B – Beiträge zur DGP-Frühjahrstagung,* 149–152.
Dilling, F. (2019b). Representation of vectors in German mathematics and physics textbooks. In S. Rezat, L. Fan, M. Hattermann, J. Schumacher, & H. Wuschke (Eds.), *Proceedings of the Third International Conference on Mathematics Textbook Research and Development* (p. 155–160). Universitätsbibliothek Paderborn.

Dilling, F. (2021). *Begründungsprozesse im Kontext von (digitalen) Medien im Mathematikunterricht. Wissensentwicklung auf der Grundlage empirischer Settings*, [Doctoral dissertation].

Filler, A. (2011). *Elementare Lineare Algebra. Linearisieren und Koordinatisieren.* Springer Spektrum.

Filler, A., & Todorova, A. D. (2012). Der Vektorbegriff. *Verschiedene Wege Zur Einführung. Mathematik Lehren, 174*, 47–51.

Kilpatrick, J. (2012). The new math as an international phenomenon. *ZDM, 44*, 563–571.

Kraus, M. E. (2016). Pfeile in der Physik: Die Schwierigkeiten beim ikonischen Modellieren. *PdN Physik in Der Schule, 65*(6), 25–32.

Malle, G. (2005). Neue Wege in der Vektorgeometrie. *Mathematik Lehren, 133*, 8–14.

Malle, G. (2005). *Schwierigkeiten Mit Vektoren. Mathematik Lehren, 133*, 16–19.

Hoche, D., Küblbeck, J., Meyer, L., Reichwald, R., Schmidt, G.-D., & Schwarz, O. (2003). *Duden Physik. Gymnasiale Oberstufe.* Duden Paetec.

Vu, D. P., Nguyen, B. V., Kraus, S. F., & Holten, K. (2020). Individuals' concepts in Physics and Mathematics Education. In S. F. Kraus & E. Krause (Eds.): *Comparison of Mathematics and Physics Education*. I. Springer Spektrum.

Lesson Plan: Combining Forces

15

Nguyen Van Bien

Lesson Title: Combining forces
Abstract: The lesson detailed below is concerned with solving situations involving two forces, using the parallelogram of forces method. The focus is on practicing this approach in different contexts. From a mathematical point of view, the applica-tion of trigonometric knowledge is particularly relevant to this lesson and the students will gain first-hand experience of vectors.

Type of school / Grade	Middle School / Grade 10
Prerequisites	– Able to combine forces in one plane and show this in a diagram – Able to resolve one force into its perpendicular components and show this in a diagram – Plan and carry out experiments to combine two concurrent forces with appropriate apparatus
Number of periods	1

N. V. Bien (✉)
Faculty of Physics, Hanoi National University of Education, Hanoi, Vietnam
e-mail: biennv@hnue.edu.vn

© The Author(s), under exclusive license to Springer Fachmedien Wiesbaden GmbH, part of Springer Nature 2022
F. Dilling and S. F. Kraus (eds.), *Comparison of Mathematics and Physics Education II*, MINTUS – Beiträge zur mathematisch-naturwissenschaftlichen Bildung, https://doi.org/10.1007/978-3-658-36415-1_15

Objectives	Mathematics – Students can apply knowledge about trigonometry and basic vector calculus Physics – Students can represent, combine, decompose, and calculate forces in a parallelogram of forc-es, using arrows

15.1 Methodical Commentary

Before the lesson, students should already have had experience with forces and their representation as arrows. In the lesson, they will solve problems involving two forces, using the parallelogram of forces method. For this purpose, the students will work on a range of tasks with illustrations as arrows, as well as digital simulations in groups. These will be discussed by the students in a plenary. Force parallelograms are a central approach in classical physics. Moreover, from a mathematics perspective, the students will also apply their knowledge of trigonometry and work with vectors for the first time.

15.2   Lesson Plan

Lesson Plan

Nr./ Time	Stage	Learning activities	Interaction form	Materials/ Resources	Methodological comments
# 1 15 min	Define the problem	*Students recall prior knowledge of forces learned in secondary school* Teacher to ask questions, students to answer individually **Questions:** *Recall the effect of a force on an object, giving examples* *Describe how to illustrate a force exerted on an object* *By observing the simulation, students identify the forces exerted on the object*	Teacher asks; individual answers	Phet simulation	These activities introduce students to the problem: how to determine the resultant force of forces exerted on the object on one plane

Nr./Time	Stage	Learning activities	Interaction form	Materials/Resources	Methodological comments
		Using Phet simulation software,[1] teachers perform simulations of three different examples of traction. The teacher then assigns tasks to the students: • Show the forces exerted on the cargo box • Describe the state of the cargo box • Illustrate the net force acting on the cargo box *Students define the problem: How to determine the combined force of forces exerted on the object on one plane* The teacher asks clarifying questions to prompt the students to show their understanding of the problem to be solved: e.g. *How is the net force of forces exerted on the object on a plane determined?* The teacher assigns the task: Determining the combined force of two concurrent forces			

Nr./ Time	Stage	Learning activities	Interaction form	Materials/ Resources	Methodological comments
# 2 30 min	Develop new knowledge	*Activity 1: Determine the combined force of two component forces (20 min)* The teacher divides the class into four to six groups to perform the tasks: • Predict the relationship between the combined force of two concurrent forces with the two component forces • Study the names and functions of the apparatus in the experimental set of the principle of combining forces • Design and execute experiments to combine two concurrent forces with given apparatus • Conclude the principle of combining two concurrent forces • Illustrate the combined force of forces exerted on the object on one plane After receiving instructions, students perform the assigned tasks on the worksheet under teacher supervision Students report and discuss the group's results The teacher gives constructive feedback The teacher sums up the material covered and the students take notes *Activity 2: Resolving one force into two perpendicular components (10 min)*	Group working	Experiment apparatus/ Worksheet	Students are encouraged to make the prediction: the combination of two concurrent forces is determined similarly by the parallelogram rule in mathematics, that is the rule of finding the sum of two vectors Students are guided to study the names and functions of the instruments / equipment provided: 02 dynamometers—measure the force, determine the direction of the force; brush—markers, draw; spring - deform when being exerted by a force; magnetic board with stand—attaching objects Students are guided to design and execute the experiment to combine two concurrent forces with given apparatus

Nr./ Time	Stage	Learning activities	Interaction form	Materials/ Resources	Methodological comments
		The teacher presents the definition and conditions of resolving a force into two component forces The teacher assigns the task: resolve the weight of an object lying on an inclined plane into two perpendicular components Instructions: • Determine the direction of the gravitational force • Apply the parallelogram rule. $\vec{P} = \vec{P_1} + \vec{P_2}$ to find $\vec{P_1}, \vec{P_2}$ in two different directions			

Nr./ Time	Stage	Learning activities	Interaction form	Materials/ Resources	Methodological comments
					Steps to be taken: S1: Tie the end of the spring (A) to the magnet base located on the board. The other end of the spring is tied to the middle of a thread. The two ends of the thread are hooked to two dynamometers placed on the board S2: Let two dynamometers simultaneously exert forces on the spring in two directions at an angle, causing the spring to lie parallel to the table surface and stretch to the O position S3: Mark on the board the projection O' of O and the direction of the two forces F_1, F_2 that the two dynamometers exert on the spring. Note down the readings of the two dynamometers S4: Use one dynamometer to pull the spring so that the spring is parallel to the table surface and also stretched to position O. Mark on the table the direction of the force $\vec{F}$ applied by the dynamometer to the rubber and note down the dynamometer readings S5: Represent the force vectors $\vec{F_1}$, $\vec{F_2}$ and $\vec{F}$ on the board and in the same chain scale. Based on the figure on the board, conclude the relationship between $\vec{F_1}$, $\vec{F_2}$ and $\vec{F}$ S6: Repeat the experiment with other pairs of forces, with different magnitudes and directions, from which to conclude the principle of combining two concurrent forces

Nr./Time	Stage	Learning activities	Interaction form	Materials/Resources	Methodological comments
					Conclusion: $\vec{F} = \vec{F_1} + \vec{F_2}$ follows the parallelogram rule. Students are guided to resolve the gravitational force into two perpendicular components

Nr./Time	Stage	Learning activities	Interaction form	Materials/Resources	Methodological comments
# 3 20 min	Apply the learning	Students solve relevant problems by using their knowledge of combining and resolving forces, working under the teacher's instruction	Individual working	Set of problems 1	
# 4 25 min	Reflection and Elaboration	Students should relate their new knowledge about forces and their representation as arrows to mathematical knowledge about vectors. To do this, they answer various reflection questions posed to them by the teacher, such as: – How can forces be described as vectors? – What is different about force arrows compared to vectors in math class? – What do the concepts have in common? – Do you know of any other physical concepts that can be described with vectors? Afterwards, students use their learned and reflected knowledge to edit further exercises (Set of problems 2)	Individual working	Questions on the left side of this table / Set of problems 2	

Tasks given to the groups:
Task 1

Problems:

Question 1: Two forces with a magnitude of 6 N and 8 N respectively are exerted on a stationery point. Draw a figure and determine the effect on the point in the following cases:

a. $\vec{F_1}, \vec{F_2}$ have the same direction	b. $\left(\vec{F_1}, \vec{F_2}\right) = 30°$
c. $\left(\vec{F_1}, \vec{F_2}\right) = 90°$	d. $\vec{F_1}, \vec{F_2}$ have the opposite direction.

Question 2: An object is lying on an inclined plane at an angle of 30° compared to the horizon. The gravitational force of the object has a magnitude of 50 N. Determine the magnitude of the components of gravitational force in the directions that are perpendicular to and parallel to the inclined plane.

Solutions:

Question 1:

a.	$F = F_1 + F_2 = 6 + 8 = 14$ (N)

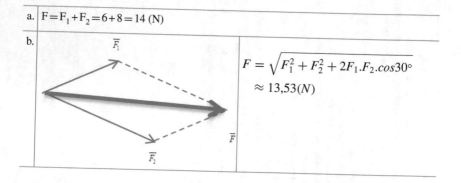

b.

$$F = \sqrt{F_1^2 + F_2^2 + 2F_1.F_2.cos30°}$$
$$\approx 13{,}53(N)$$

c. 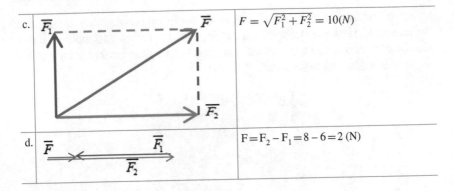	$F = \sqrt{F_1^2 + F_2^2} = 10(N)$
d.	$F = F_2 - F_1 = 8 - 6 = 2$ (N)

Question 2:

$P = 50$ N $\alpha = 30°$

$$P_2 = P \cdot \cos 30° = 25\sqrt{3}(N)$$

$$P_1 = P \cdot \sin 30° = 25(N)$$

Task 2

Problems:

Question 1: An object is hung on a rope, lying balanced, as shown in the picture. Determine the object's mass, given that the tension of each end of the wire is 200 N and the angle between the two tensions is 150°.

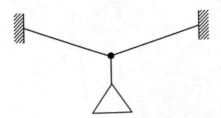

Question 2: A mirror hangs on a wall, as shown in the picture. Considering that the mass of the string is negligible, the string is not stretched. Determine the tension of each end of the rope, given that the mass of the mirror is 2 kg and the angle between two tensions is 60°.

Question 3: Your task is to weigh an object weighing between 11 and 20 (N) using only one dynamometer with a measuring limit of 10 N and a length of thin, light, non-stretch string. Explain how you would complete the task.

Solution:

Question 1:

$$P = F = \sqrt{F_1^2 + F_2^2 + 2F_1F_2cos150°} \approx 103,5(N)$$

The mass of the object. $m = \frac{P}{g} \approx 10,56(kg)$.

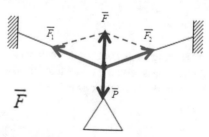

Question 2:

$F_1 = F_2, \left(\overrightarrow{F_1}, \overrightarrow{F_2}\right) = 60°$

$$P = F = \sqrt{F_1^2 + F_2^2 + 2F_1F_2cos60^\circ} = m.g = 2 \cdot 9{,}8 = 19{,}6(N)$$

$$\Leftrightarrow \sqrt{3}F_1 = 19{,}6 \Leftrightarrow F_1 = 11{,}3(N)$$

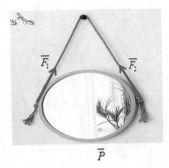

Question 3: From question 1 and 2 we have a methodology: hanging a heavy object on the two sections of the rope, and tying the dynamometer to a section of the rope ensuring that the angle between the two wires is an acute angle.

Lesson Plan: Basic Concepts Related to Vectors

16

Le Tuan Anh

Lesson Title: Basic concepts related to vectors
Abstract: In the lesson presented here, students learn about fundamental concepts related to vectors. To do so, they primarily work at the qualitative level with vectors represented as directed segments. Another focus of the lesson is the application of the concepts to physics, especially to the description of forces in mechanics.

Type of school / Grade	High School / Grade 10
Prerequisites	– Basic knowledge of forces from physics
Number of periods	1
Objectives	Mathematics: – Students can identify the concepts of vector, equal vector, opposite vector, zero vector, and other related concepts – Students can discuss and work in small groups effectively – Students can implement critical thinking skills and find strategies to solve problems Physics: – Students can describe phenomena in the natural sciences, including physics, using vectors and use vectors to explain some phenomena related to physics

L. T. Anh (✉)
Faculty of Mathematics and Informatics,
Hanoi National University of Education, Hanoi, Vietnam
e-mail: anhlt@hnue.edu.vn

© The Author(s), under exclusive license to Springer Fachmedien Wiesbaden GmbH, part of Springer Nature 2022
F. Dilling and S. F. Kraus (eds.), *Comparison of Mathematics and Physics Education II*, MINTUS – Beiträge zur mathematisch-naturwissenschaftlichen Bildung, https://doi.org/10.1007/978-3-658-36415-1_16

16.1 Methodical Commentary

The goals of the lesson are for students to learn and apply the concept of vectors. A description of forces in mechanics serves as motivation. During the development and practice of the concept, students mostly work in groups, but there is also individual work and some parts of the lesson are moderated by the teacher. Much of the work is based on worksheets created specifically for the lesson.

16.2   Lesson Plan

Nr./Time	Stage	Learning activities	Interaction form	Materials/Resources	Methodological remarks
# 1 12 min	Motivation	– Students review types of force, how to represent a force, the gravity of two similar objects placed on the same plane, and the equilibrium of two forces	Group work	– Worksheet 1/High school Physics or Science textbook	– Students have learned about forces, gravity, and the equilibrium of two forces. Teaching concepts related to vectors should start with the prior knowledge and experience of students – Teachers should have assigned students tasks in the previous lesson that will allow them to study this content effectively – From the types of force, the teacher helps students understand the concept of vector. From the gravity of two similar objects placed on a plane, students grasp the concept of equal vectors. From the equilibrium of two forces, students understand the concept of opposite vectors – The teacher can have one group present their answers to Worksheet 1. Alternatively, each group can present their part of the worksheet and other groups can comment or supplement
# 2 48 min	Approaching new knowledge	*The concept of vector* Students read the definition of vector in the textbook and identify the following items: vector, initial point, terminal point; how to draw a vector; the notation of a vector	Teacher asks, students answer/Individual work	Mathematics textbook/Worksheet 2	– There are four different definitions of a vector on a plane: an equivalent class of ordered pairs of points, an equivalent class of directed segments, an element of a vector space, and a directed segment. From a mathematical viewpoint, the two first definitions are accurate and clear. However, they are not suitable for application to Vietnamese physics and mathematics classes (e.g., force, velocity, congruent triangles and circles). The third definition often appears in mathematics books in higher education. For these reasons, the fourth definition is usually used in Vietnamese high school mathematics textbooks and classes Given the segment *AB, if we consider A* the initial point and *B* the terminal point, then the segment has a direction from *A* to *B*. We can say that *AB* is a directed segment *Definition: A vector is a directed segment* – Students need to demonstrate the concept of vector through concrete examples. This helps to reinforce the concept

Nr./Time	Stage	Learning activities	Interaction form	Materials/Resources	Methodological remarks
		Equal vectors Students identify parallel vectors, vectors with the same direction or opposite directions, and the magnitude of a vector From the representation of the gravity of two similar objects, the teacher helps students understand the concept of equal vectors	Group work	Mathematics textbook/ Worksheets 3 & 4	Students become familiar with the concept of equal vectors from a representation of the gravity of two similar objects placed on a plane. Note that the direction of gravity is always down-ward (to words the Earth), i.e. perpendicular to the plane perpendicular Worksheet 3 helps students to understand and identify parallel vectors, vectors with the same direction, and vectors with opposite directions Worksheet 4 helps students to identify equal vectors
		Opposite vectors and zero vector From the representation of the equilibrium of two forces, the teacher presents the concept of opposite vectors Then, the teacher presents the concept of zero vector	Group work/ teacher presents	Mathematics textbook/ Worksheet 5	Students are shown a representation of the equilibrium of two forces to illustrate the concept of opposite vectors Note that, when representing the equilibrium of two forces in physics, the initial point of each force is considered, but in mathematics, we do not need to consider the initial points of opposite vectors. Teachers should help students identify this difference
# 3 20 min	Practice	Consolidate learned concepts	Individual work/ group work/ whole-class disscussion	Textbook/ Worksheets 6 & 7	Worksheet 6 is for individual work and Worksheet 7 is for group work
# 4 10 min	Application	Instruct students to relate and apply concepts to phenomena in physics and the natural sciences	Group work/Individual work	Worksheet 8/ documents about physics and the natural sciences/ Internet	This activity helps students relate learned knowledge to situations in physics and the natural sciences Reiterate that, in physics, we pay attention to the initial points of forces

Worksheets:

Worksheet 1

(i)   List some types of force that you know or have learned:

(ii) How do you represent a force?

(iii) What are the direction and intensity of the gravity $\overline{P}$ and $\overrightarrow{P_1}$ of two similar objects placed on a table (Figure 1)?

(iv) What are the direction and intensity of the two forces in equilibrium $\overline{N}$ and $\overline{P}$ (Figure 2)?

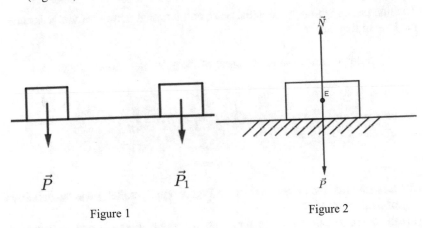

Figure 1   Figure 2

Worksheet 2

(i)   Read about the concept of vector in the textbook. Explain the following points: vector; initial point, terminal point; how to draw a vector; the notation of a vector.

(ii) In Figure 3, which images represent vector $\overrightarrow{CD}$?

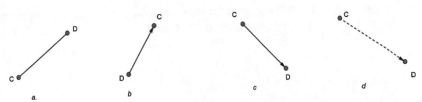

Figure 3

(iii) Given two points E and F, draw vector $\overrightarrow{EF}$.

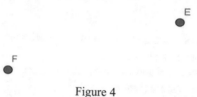

Figure 4

Worksheet 3

The line passing through the initial point and terminal point of a vector is called the *base* of that vector.

(i) What do you notice about the bases of the pairs of vectors in Figure 5?

Figure 5

(ii) It is said that the pairs of vectors in Figure 5 are parallel. State the definition of parallel vectors.

(iii) Read the definitions of vectors with the same direction and vectors with opposite directions. Identify in Figure 5:
- Vectors with the same direction:
- Vectors with opposite directions:

Worksheet 4

The length of segment AB is called the magnitude of vector $\overrightarrow{AB}$.

(i) Suppose that each small square in Figure 6 has a side length of 1. Draw three vectors whose respective magnitudes are 3, 5, and $\sqrt{41}$.

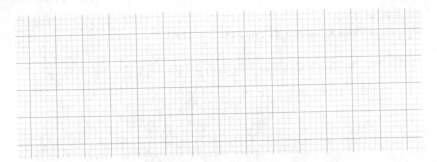

Figure 6

(ii) Figure 7 represents the gravity of two similar objects placed on a table.
  • Describe the direction and length of the two vectors.
  • These vectors are called equal vectors. Define equal vector.

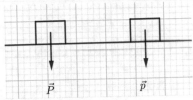

Figure 7

(iii) Identify vectors that are equal to vector $\overrightarrow{AB}$ in Figure 8:

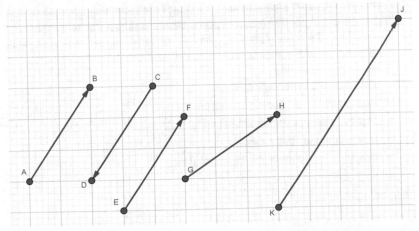

Figure 8

Worksheet 5

Figure 9 represents the equilibrium of two forces.

(i) Describe the direction and magnitude of the two vectors in Figure 9.

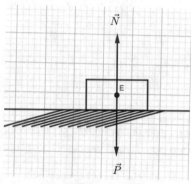

Figure 9

(ii) Vector $\vec{P}$ is called the opposite vector of vector $\vec{N}$. This is denoted $\vec{P}=-\vec{N}$.

State the definition of opposite vectors.

Worksheet 6

In the examples below, we consider nonzero vectors.

(i) Given two distinct points M and N, identify vectors that have M or N as either the initial point or terminal point. Denote and name those vectors.

(ii) Given three distinct points P, Q, and R, identify vectors that have initial points and terminal points belonging to the set {P, Q, R}

(iii) Show or draw a vector:

- that is parallel to vector $\overrightarrow{EF}$;
- that has the same direction as vector $\overrightarrow{EF}$;
- that has the opposite direction to vector $\overrightarrow{EF}$;
- that is equal to vector $\overrightarrow{EF}$;
- that is the opposite vector of vector $\overrightarrow{EF}$.

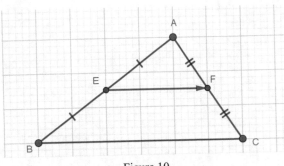

Figure 10

Worksheet 7
Prove that three points A, B, C are colinear if and only if the two vectors AB and AC are parallel.

Worksheet 8
Describe situations in natural science or physics that correspond with the following concepts:

(i)   Vector;
(ii) Two vectors with the same direction; two vectors with opposite directions;
(iii) Equal vectors;
(iv) Opposite vectors.

Comparison: Differential Calculus Through Applications

17

Gero Stoffels, Ingo Witzke and Kathrin Holten

17.1 Introduction

Three perspectives on the concept of derivatives at school are commonly applied: the geometric, the algebraic–analytical and the application-oriented (Danckwerts & Vogel, 2006, p. 85). The geometric perspective is the classical approach used by Leibniz, where the slope of the tangent at a point in the graph of a function is identified with the limit of the difference quotient of a function at that point (Witzke, 2014, pp. 27–28). From an algebraic–analytical perspective based on the axiomatization of real numbers as Dedekind-complete ordered fields, one definition states: "if the function f is differentiable at x_0, $f'(x_0)$ is the slope of the tangent t of f at x_0," so it is actually the other way around. Often the geometric perspective is used for introducing the concepts of calculus on given curves as graphs of a function. However, students' experience with tangents as global support tangent lines, e.g., tangents to circles, differs from the new interpretation of a tangent as a straight line that locally has the same slope at the point of osculation as the function at the same point (Danckwerts & Vogel, 2006, pp. 45–46). These different perspectives on tangents are a well-described problem for students starting

G. Stoffels · I. Witzke · K. Holten (✉)
Mathematics Education, University of Siegen, Siegen, Germany
e-mail: holten@mathematik.uni-siegen.de

G. Stoffels
e-mail: stoffels@mathematik.uni-siegen.de

I. Witzke
e-mail: witzke@mathematik.uni-siegen.de

F. Dilling and S. F. Kraus (eds.), *Comparison of Mathematics and Physics Education II*, MINTUS – Beiträge zur mathematisch-naturwissenschaftlichen Bildung, https://doi.org/10.1007/978-3-658-36415-1_17

on differential calculus. The algebraic–analytical perspective seems to be more complicated than the geometric perspective, as it deals with the concept of limit, which is not mandatory in all curricula in its rigorous form, at least in German classrooms today. For this reason, the formal definition of the differential quotient in x_0

$$f'(x_0) = \lim_{x \to x_0} \frac{f(x) - f(x_0)}{x - x_0}$$

is often interpreted in a geometric way as the slope of the secants approaching the tangent of f at x_0. As shown in Fig. 17.1, there is another epistemological problem for students in understanding the differentiation process as approximation and not as coincidence of points (Danckwerts & Vogel, 2006, p. 50).

Similar observations can be made with the concept of integration.

These problems of the geometric and algebraic–analytical concepts may be avoided if the third approach, the application-oriented, receives more attention through the use of calculus in mathematics and physics.

Not only can this shift of focus be productive but also the connection of mathematics and physics plays a major role. For example, the first problem mentioned above does not have to be considered if teachers use physical contexts. Physics can also address the second problem, because measurements errors in measurements are only possible in a discrete manner. As such, the concept of the rate of change can be introduced by means of velocity. From a theoretical point of view, the rate of change is a continuous quantity, whereas from an application-oriented perspective the measured velocity is a discrete quantity because of the restrictions of the measuring process. The concrete idea of instantaneous velocity, which is a theoretical concept, as a limit of average velocity, which is empirically measurable, may help to bridge the epistemological gap between difference quotient and differential quotient or between average rate of change and instantaneous rate of change in general. Studies have shown that few pupils associate derivatives with rates of change or instantaneous velocity, although "vast efforts of subject matter didactics have been made in the last decades to introduce concepts of calculus in application contexts" (Witzke & Spies, 2016, p. 147). However, it is also challenging to present a physical context in mathematics classes in an authentic way. Dilling and Krause (2020) analyzed German textbooks regarding authentic context from kinematics. They found that teachers without knowledge in physics could not present the kinematic context just by using the textbook.

In Vietnam, the derivative concept is introduced in the 11th-grade mathematics program (Tran and Vu, 2011). To approach the derivative concept, the textbook poses two physics problems: calculating instantaneous velocity and finding

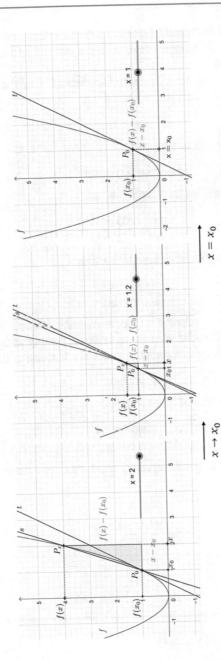

Fig. 17.1 Epistemological problem as a result of illustrating the limit process as approximating the slope of the tangent with the slope of a secant of f

instantaneous intensity of current. These are two examples of the application-oriented approach. The first example deals with the concept of motion, which we have already mentioned:

An object moves along the axis s'Os. The distance s of the movement is a function of time t (Fig. 17.2).

$$s = s(t)$$

In the interval t_0 to t the object has gone:

$$s - s_0 = s(t) - s(t_0)$$

If the object is moving uniformly, the proportion

$$\frac{s - s_0}{t - t_0} = \frac{s(t) - s(t_0)}{t - t_0}$$

is a constant for every t. The finite limit (if it exists)

$$\lim_{t \to t_0} \frac{s(t) - s(t_0)}{t - t_0}$$

is called "the instantaneous velocity" of the movement at time t_0.

The second example from the Vietnamese textbook, dealing with physics, states:
The amount of electricity transferred in a wire is a function of time t:

$$Q = Q(t)$$

erage amperage over a period of time $|t - t_0|$ is:

$$I_{tb} = \frac{Q(t) - Q(t_0)}{t - t_0}$$

If $|t - t_0|$ is smaller, this ratio will represent more accurately the amperage at t. It defined thus:

The finite limit (if it exists)

$$\lim_{t \to t_0} \frac{Q(t) - Q(t_0)}{t - t_0}$$

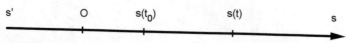

Fig. 17.2 Axis s'Os (Tran and Vu, 2011)

is called the instantaneous amperage of the current at time t_0.

In the Vietnamese textbook (Chu & Nguyen, 2017), the concept of derivative is given inductively as the finite limit of proportion as:

$$\frac{f(x) - f(x_0)}{x - x_0}$$

when x goes to x_0. $f(x)$ is a function defined in the interval $(a; b)$ with $x_0 \in (a; b)$. Earlier, we called this approach the algebraic–analytical perspective. The geometric meaning of derivative is introduced as the slope of a tangent to the graph of function $f(x)$ at point $(x_0; f(x_0))$ —the geometric perspective.

Although teachers in Vietnam, as in Germany, try to implement physical applications in mathematics lessons in the aforementioned ways, from a mathematics educator's point of view, more work has to be done in this setting, particularly regarding the findings of Witzke and Spies (2016, p. 147). In which ways can contexts applying calculus, e.g., dynamics, enrich learning environments to initiate a sustainable idea of derivative? If, for example, units are covered in mathematics classes, physics could help to arrange the concept of motion in a physically correct way (see chapter "Comparison: Equations"). A physics educator could ask: Which mathematical content and skills could help to arrange the concept of motion? In the following, some educational research topics referring to (physical) applications of differential calculus are presented to prepare an interdisciplinary theoretical basis for concrete lesson plans (cf. lesson plans "Differential calculus through applications" and "Capacitor charge and discharge process") to address questions like those mentioned above.

17.2 Historical evolution of differential calculus and its connection to physics

Differential and integral calculus were developed in the seventeenth century by Newton and Leibniz relatively independently of each other. There have been a number of approaches and results since antiquity, both for differential calculus and integral calculus (in the sense of determining the slopes of tangents of a curve and the content of curved surfaces and bodies). Famous examples of this were provided by Archimedes (287–212 BC), "the greatest mathematician of ancient times" (Edwards, 1979, p. 29), who determined the tangents to the spiral named after him, as well the area of parabolic segments. Until the seventeenth century, however, only specific curves, surfaces, or solids had been examined, and no general methods had been developed that would be suitable for solving classes of problems in a uniform and systematic way. This only changed with the discovery and use of so-called infinitesimal quantities.

A principle that was discovered by Cavalieri in 1635 and that allows the content of geometric objects to be compared—without calculation, especially without integral calculus—is still used in today's school lessons. It reads as follows (cf. GW Evans, 1917):

> If between the same parallels any two plane figures are constructed, and if in them, any straight lines being drawn equidistant from the parallels, the included portions of any one of these lines are equal, the plane figures are also equal to one another; and if between the same parallel planes any solid figures are constructed, and if in them, any planes being drawn equidistant from the parallel planes, the included plane figures out of any one of the planes so drawn are equal, the solid figures are likewise equal to one another.

Cavalieri's principle allows the deduction of the contents of figures and bodies from the indivisibles. A classic application is to prove that all cones with the same base area and height have the same volume. It is noteworthy that the Cavalierian principle is a purely geometric principle: the geometric properties of planar or spatial figures are compared and the same content is inferred. There is no calculation and no algebra used.

Leibniz's merit is that he took up the idea of infinitely small quantities and combined different approaches, including those of Fermat and Pascal, into a consistent theory, which he called *calculus differentiali* and *calculus integralis*. The term "calculus" has been preserved in English to this day as an expression that denotes analysis. In German, the word *Infinitesimalrechnung* indicates the origin of analysis.

The first printed textbook on differential and integral calculus was the *Analyse des Infiniment Petits pour l'Intelligence des Lignes Courbes* by the Marquis de l'Hôpital, published in 1696. He was a wealthy French nobleman, a marquis who was interested in mathematics and was introduced to the secrets of calculus by Johann Bernoulli—for a handsome fee. The manuscript written by Bernoulli in 1691–1692 was only discovered in 1922 and the actual author of the analysis identified. Bernoulli presented Leibniz's calculus in a systematic way based on its didactic intentions. The object of the calculus are curves that were imagined to be constructed in the classical Greek way. Following an idea of Descartes, these curves were then embedded in a coordinate system and described by an equation. Such an equation, e.g., $y = x^2$, describes the relationship between two geometric quantities, more precisely distances: between the abscissa x of a point of the curve and its ordinate y. At this point, the calculus begins by first devising rules on how to set up the differentials dx and dy of the geometric quantities x and y with respect to a given curve. In the example of the parabola, these rules state that $dy = 2dx$. In Leibniz's calculus, theorems are formulated, e.g., that dy/dx indicates the slope of the tangent at the point with the abscissa x and ordinate y. In the example of the parabola above, the slope is $\frac{dy}{dx} = 2x$. In his textbook, Bernoulli made it clear that curves can also be given in other ways than geometric construction on a drawing sheet. Examples of this are the catenary and the brachistochrone.[1] At that time, mathematics and physics were not yet separate disciplines, but rather one. This is also very clearly expressed by the title of Newton's work, in which he presents his theory of fluxions and fluents for the treatment of mechanical problems: *Philosophiae Naturalis Principia Mathematica*. Newton develops his mathematical theory in the context of mechanics and sees both as *Philosophia Naturalis*—differential and integral calculus by applications, as we would say today.

What is the meaning of differentials and fluxions, i.e., infinitesimal quantities? This has been debated for over a century. Leibniz himself illustrated this with a "characteristic triangle." (Fig. 17.3)

Leibniz assigned a right-angled triangle to each point C of a curve, the catheti of which are the differentials dx and dy and the hypotenuse of which is the differential of the arc length ds. He used the integral ds to determine the arc length. This has led to the idea that the characteristic triangle has its hypotenuse in common with the curve, i.e., the tangent at point C not only touches the curve at point C but also has the infinitesimal distance ds in common with the curve. According to Leibniz, the characteristic triangle is similar to the tangent triangle, which consists of the tangent section t, the subtangent section m, and the ordinate y of the point C, i.e., $ds : dx : dy = t : m : y$ applies in the calculus. From this law, it follows immediately that the quotient of dy/dx indicates the slope of the tangent, because it is $dy : dx = y : m$.

Contemporaries of Leibniz and Newton were enthusiastic about the complex problems that could be solved with the new calculus. Galileo's statement that the book of nature was written in the language of mathematics was irrefutably proven by Newton's *Principia*. The downside was the conceptual ambiguity that was attached to the concept of infinitesimal size. A sharp criticism came in 1734

---

[1] Catenary: Which curve describes a chain freely suspended at its ends? While Galileo still suspected that it was a parabola, Bernoulli showed that the curve can be represented by the hyperbolic cosine. Brachistochrone: In a vertical plane, two points, A and B, are given. Which curve describes the path of a ball that rolls from A to B in the shortest time possible?

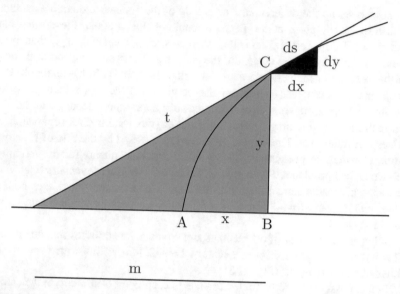

Fig. 17.3 Leibniz's characteristic triangle (black) and the tangent triangle (gray) are "similar"

from Bishop G. Berkeley, who mocked infinitely small quantities as "ghosts of departed quantities" in a polemic that was widely noticed at the time. Despite all the criticism, Berkeley also had to admit that the calculus (and its competitor, Newton's fluxion calculus) were extremely successful. They had become irreplaceable in the natural sciences.

In the following two centuries, one can summarize the historical development: the aim was to clear up this conceptual ambiguity. Euler had the idea that instead of starting from curves and describing them afterward with an equation, the procedure could be reversed, i.e., starting from equations and assigning curves to them afterward. Since equations are defined with numbers, infinitely small quantities may become obsolete in this way.

In 1748, Euler's *Introductio in Analysin Infinitorum* was published, a textbook in which not a single curve is depicted, in contrast to L'Hôpital's *Analyse des Infiniment Petits pour l'Intelligence des Lignes Courbes*, which had appeared only half a century earlier. The subject of Euler's analysis are functions to which curves can be assigned via a graph, but no longer need to be. Euler's concept of function was not yet as elaborate as the modern one, but it has essential characteristics from it. Euler saw functions as fundamental for analysis not only because they can be used to avoid infinitely small quantities but also because they make it possible to describe physical phenomena appropriately, and thus he was completely in the tradition of Newton. In his collected works, 29 volumes focus on mathematical subjects and 45 on physical topics. Already, these numbers show the close relationship between mathematics and physics for Euler. This can be definitely seen as a feature of its time. Lagrange, d'Alembert, and Laplace also achieved excellent results in both mathematics physics. The two areas were seen by them as a common field of work.

The second step in the endeavor to make infinitely small quantities superfluous to analysis was the invention of the concept of limit. The key idea was published by Cauchy in his *Cours d'Analyse* from the year 1821. If the values of a variable, he writes, approach a fixed value so "that they differ from this value as little as one wants, then the fixed value is called the limit of all other values." The term "limes" also goes back to Cauchy. In the second half of the nineteenth century, Weierstrass was finally able to describe "approaching a variable" purely arithmetically with the help of its well-known ε–δ definition. He gave the analysis of real-valued functions the shape that is still valid today.

Until the dawn of modern times, mathematics and physics were regarded as a single science. This changed with Hilbert's formalistic conception of mathematics: Mathematics has become an independent discipline that works without ontological ties. Nevertheless, the two areas still fertilize each other today. Analysis would not have been developed without physics, and physics could not have been formulated without mathematics. To this day, it can be said that physics is the prime example of mathematical applications and the stimulus for mathematical theories.

17.3 Differential calculus for interdisciplinary teaching and learning

The connection of mathematics and physics already shown in the historical evolution of differential calculus can also be supportive in educational contexts—especially interdisciplinary ones. A premise for the supportive characteristics of such interdisciplinary settings is the awareness of limitations as well as benefits for each subject. In the following, we discuss these limitations and benefits.

The most common content that connects differential calculus and physics, especially in undergraduate mathematics and physics, is dynamics. Many mathematical textbooks use the concept of instantaneous velocity as a standard example of the instantaneous rate of change (Witzke, 2014, p. 27; Dilling & Krause,

2020). Dynamics in physics marks the beginning of a more complex mathematized physics instruction (Martínez-Torregrosam et al., 2006). Other physical applications of differential calculus, e.g., alternating current, the induction principle, or radioactive shielding, exist as well. They also offer educative opportunities for interdisciplinary teaching and learning. However, because of the common approach, this reflection focuses on dynamics.

The following sections give an overview of how differential calculus is used in mathematics and physics at school, in which ways mathematics education discusses physical applications in mathematical learning processes, and in which ways physics education deals with the mathematical methods of differential calculus.

17.3.1 Differential calculus in mathematics and physics at school and university

Rasmussen et al. (2014) stated that many students enroll in calculus each year. It depends on the country's education system if the courses take place at the tertiary or secondary level. There may be a possibility that there are differences if calculus is learned in tertiary or secondary education, but Rasmussen et al. did not think that the differences were in the content, but more regarding cultural or institutional factors. Bressoud et al. (2016) gave an overview of the teaching of calculus in France, Germany, Uruguay, Singapore, South Korea, Hong Kong, and the US, focusing on differences among their systems. All countries provided calculus teaching in higher secondary education except the US, though there it is at least common to take a calculus course in high school in preparation for university. Bressoud et al., (2016, p. 16) made this point very explicit in stating: "Calculus in the United States is and always has been considered a university-level course. Curiously, it is now predominantly taught in high school." This situation goes back to a decision of the College Board, which established the Advanced Placement Program. This is interesting in respect of research on educational implications of teaching and learning calculus insofar as the run of students on these courses led to a calculus reform in the 1990s with a focus on fostering conceptual understanding, rather than schematic routine learning in differential calculus courses (Bressoud et al., 2016).

In general, a wider spectrum of mathematical activities in differential calculus has found its way into modern curricula in Germany and Vietnam. The general guidelines (standards) require of all schools in Germany (*Bildungsstandards*) teaching calculus courses at the higher secondary level that students be qualified to use functions for describing quantizable relations, compose functions, interpret the derivative as the local rate of change, describe local rates of change as functions, determine derivatives of functions using "rules for derivations," and construct a graph of the derivative of a function from the graph of the function. Similar requirements are also given for dealing with integrals. This shows that in recent curricula, multiple views on calculus have been addressed. A fruitful integration of calculus in mathematics and science can be found explicitly only in Singapore, where "calculus has to begin early to support physics learning" (Bressoud et al., 2016, p. 20).

Differential calculus in physics is usually introduced in the context of kinematics. A lot of research in the physics-education community has been done since the 1980s (Trowbridge & Dermott, 1980, 1981; Beichner, 1994) in this field, showing that students lack basic knowledge in calculus, especially when starting with mechanics. More recent studies have show, that these difficulties persist even when students are introduced to these concepts at the university level (Christensen & Thompson, 2012).

17.3.2   Physical applications in mathematics teaching and learning from a mathematics-education perspective

From a mathematics-education researcher's perspective, a lot of general issues of mathematics education can be addressed by looking at the field of differential calculus in particular (Bressoud et al., 2016). This subsumes developing general theories of student learning (Tall, 2009), beliefs of teachers or students (Witzke & Spies, 2016), conceptual change processes regarding content-specific concepts like limits, derivatives, and infinitesimals (Greefrath et al., 2016; Tall & Vinner, 1981), or task construction (Christensen & Thompson, 2012). Rasmussen et al. (2014) and Bressoud et al. (2016) have provided good overviews of the different research approaches and their interdependence.

In Bressoud et al. (2016), the connection to physics is notable only in subsections dealing with the curricula of different countries. In Rasmussen et al. (2014), the connection is a side note on branches of theory development in the context of research in the teaching and learning of calculus. According to these authors, one direction of recent research lays in "crossing disciplinary boundaries to physics" (Rasmussen et al., 2014, p. 508). One of the articles mentioned in Christensen and Thompson (2012) will be discussed more deeply in 0, as the authors have a physics-education background. The other mentioned article discusses the question of how students use physics to reason about calculus task, in which Marrongelle (2004) gives some examples of how physics can support learning differential calculus. Her approach differs strongly from the usual perspectives on context tasks, because she focuses on students' insertion of contexts for solving mathematical tasks, rather than their resolving the physics in application-oriented tasks for their mathematical solving processes. The major result of her case study is the Physics Use Classification Scheme (PUCS), which was inspired by Zandieh's (2000) framework for analyzing students' understanding of derivatives, where the dimension "representation or context" also includes paradigmatic physical contexts like speed or velocity. Marrongelle (2004, p. 263) provided with her PUCS a categorization scheme describing in which ways students use physical contexts. She distinguishes four user types: the "contextualizer," who introduces a physical context for solving a task, the "example user," who uses examples for grasping concepts of calculus, the "language mixer," who uses concepts of physics and mathematics, though it is not clear into which other category this user type should be subsumed, and the "nonuser," who does not use any physical context during problem-solving tasks. These informed her multiple-case study on participants in a project for curriculum development interlacing the mathematics and physics curricula. Despite this context, Marrongelle did not infer that any of the user types primarily reflected the aims of the curriculum, but that there were different modes of thinking and multimodal possibilities dealing with calculus tasks. In her article, there is also no evidence that any of the identified user types is more successful in solving differential calculus tasks, but at least physics seems to be supportive for some students learning differential calculus, e.g., "contextualizers" or "example users."

17.3.3 Mathematical methods of differential calculus in physics teaching and learning from a physics-education perspective

In early physics-education research, there were some studies indicating problems in kinematics that were connected to concepts and methods of differentiation and integration (Trowbridge & Dermott, 1980). In general, there seems to be no doubt in the physics-education community that the mathematical methods of differential calculus are irreplaceable for doing physics on a more complex level (Martínez-Torregrosa et al., 2006; Rizcallah, 2018).

A more recent study tried to identify if problems in physical tasks originate in the mathematical concepts and methods themselves or if they can be traced back to the physical contexts. To answering this question, Christensen and Thompson (2012, p. 1) developed the idea of stripping physics from physics tasks, resulting in "physics-less physics questions." With this construction, they showed that first- and third-semester students were not able to solve such constructed "mathematical tasks." In their conclusions, they gave the interpretation that the "type of mathematical tasks we want our students to do in a physics class may simply be foreign to their mathematical ways of thinking" (Christensen & Thompson, 2012, p. 5), which can be seen as an evidence for separate "domains of subjective experiences" (Bauersfeld, 1994). The major result of this study was that physics-education researchers would probably interpret the mistakes and problems of the students on such exercises in a kinematic context in terms of the difficulty of these contexts, where Christensen and Thompsen (2012, p. 5) claim, that "this physics-less physics question points to at least a few students who use notions about area to answer questions about derivatives independent of physical context." Nevertheless the transfer from mathematics to physics is a research field in physics education. For example, Cui et al. (2006) focused on the transfer from calculus to physics.

A more subject-matter didactic approach in physics education proposes that there is indeed a problem in interpretation of the mathematical methods of differential calculus, especially the modern perspective based on Cauchy's theory on limits. Martínez-Torregrosa et al., (2006, p. 448) referred to their previous research in illustrating the problem, which showed that only a few students who "use differential calculus in physics fully understand what they are doing." The same applied for 88% of teachers in high schools who participated in their studies. By referring to results of mathematics-education research Martínez-Torregrosa et al. identified the main problem in the proper formulation of the concepts of differential calculus. Their proposed solution to the problem was based on a historical investigation of different views on calculus, from Leibniz to Cauchy and finally to Fréchet, who formulated a definition of the differential of a multivariable function that approximates the function by a linear function whose increments can be interpreted as approximation errors. For this article, the concrete definition of the differential is not as important as the reasons that Martínez-Torregrosa et al. saw benefits in using Fréchet's rather than Cauchy's definition of differentials:

> We believe that this concept of differential may bring about the necessary reconciliation of physical usefulness with mathematical rigor and accuracy, because its definition is as precise as Cauchy's but it has a clear meaning linked to the idea of approximation, which is so important in physics since the origin of calculus.
> (Martínez-Torregrosa et al., 2006)

This shows how necessary the discourse of mathematics education and physics education is for identifying appropriate concepts of differential calculus not hindering but supporting mathematical and physical learning.

17.4 Combining approaches from an interdisciplinary perspective of mathematics and physics education

Recent studies have shown that virtually no students link "derivative" with the application-oriented interpretation of "rate of change" or "concrete examples of real-life applications like instantaneous velocity" (Witzke & Spies, 2016, p. 147). This is especially remarkable because the historical roots of differential calculus are to be found in the analysis of geometric curves, and also in Newton's theory of mechanics.

Many curricula already demonstrate a deeper interconnection of mathematics and physics, especially in the context of differential calculus. There are already some results from physics-education and mathematics-education research crossing the border between mathematics and physics. These results show that on the one hand, for some students, explicit connections between mathematics and physics can foster their learning of concepts and methods of differential calculus in general, while on the other hand, the use of mathematical methods and concepts of differential calculus is quite different in physics. As a teacher, one should be aware of these challenges in both subjects.

It seems that there is still a lot of potential for (empirical) research describing in which ways physics contexts can support mathematical learning and how we can improve the learning of differential calculus so as to be more appropriate for physics contexts, or as Rasmussen et al. (2014) say: "An area of even greater need, however, is that of the relationship between calculus and the client disciplines of engineering, physics, biology, and chemistry." Obviously, it seems productive to see physics- and mathematics-education researchers as partners paving the way to a broader picture of learning differential calculus.

References

Bauersfeld, H. (1994). Theoretical perspectives on interaction in the mathematics classroom. In R. Biehler, R. W. Scholz, R. Sträßer, & B. Winkelmann (Eds.), *Didactics of mathematics as a scientific discipline* (pp. 133–146). Kluwer.

Beichner, R. J. (1994). Testing student interpretation of kinematics graphs. *American Journal of Physics, 62*, 750.

Bressoud, D., Ghedamsi, I., Martinez-Luaces, V., & Törner, G. (2016). *Teaching and learning of calculus.* Springer. https://doi.org/10.1007/978-3-319-32975-8.

Christensen, W. M., & Thompson, J. R. (2012). Investigating graphical representations of slope and derivative without a physics context. *Physical Review Physics Education Research, 8*(2). https://doi.org/10.1103/PhysRevSTPER.8.023101.

Chu, C. T., & Nguyen, T. D. (2017). Analysis of didactic transposition in teaching the concept of derivative in high schools in the case of Vietnamese textbooks in 2000, 2006 and American textbook in 2010. *HNUE Journal of Science Educational Science, 62*(6), 10–18. https://doi.org/10.18173/2354-1075.2017-0123.

Cui, L., Rebello, N. S., & Bennett, A. G. (2006). College students' transfer from calculus to physics. *AIP Conference Proceedings, 818*(1), 37–40. https://doi.org/10.1063/1.2177017.

Danckwerts, R., & Vogel, D. (2006). *Analysis verständlich unterrichten: Mathematik Primar- und Sekundarstufe.* Elsevier.

Dilling, F., & Krause, E. (2020). Zur Authentizität kinematischer Zusammenhänge in der Differentialrechnung—eine Analyse ausgewählter Aufgaben. *MNU Journal, 2*, 163–168.

Edwards, C. H., Jr. (1979). *The historical development of the calculus.* Springer.

Evans, G. W. (1917). Cavalieri's theorem in his own words. *American Mathematical Monthly, 24*, 447–451.

Greefrath, G., Oldenburg, R., Siller, H. S., Ulm, V., & Weigand, H. G. (2016). Aspects and "Grundvorstellungen" of the concepts of derivative and integral. *Journal Für Mathematik-Didaktik, 37*(S1), 99–129. https://doi.org/10.1007/s13138-016-0100-x.

Martínez-Torregrosa, J., López-Gay, R., & Gras-Martí, A. (2006). Mathematics in physics education: Scanning historical evolution of the differential to find a more appropriate model for teaching differential calculus in physics. *Science & Education, 15*(5), 447–462. https://doi.org/10.1007/s11191-005-0258-y.

Marrongelle, K. A. (2004). How students use physics to reason about calculus tasks. *School Science and Mathematics, 104*(6), 258–272. https://doi.org/10.1111/j.1949-8594.2004.tb17997.x.

Rasmussen, C., Marrongelle, K., & Borba, M. C. (2014). Research on calculus: What do we know and where do we need to go? *ZDM Mathematics Education, 46*(4), 507–515. https://doi.org/10.1007/s11858-014-0615-x.

Rizcallah, J. A. (2018). Projectile motion without calculus. *Physics Education, 53*(4), 045002.

Tall, D. O. (2009). Dynamic mathematics and the blending of knowledge structures in the calculus. *ZDM Mathematics Education, 41*(4), 481–492. https://doi.org/10.1007/s11858-009-0192-6.

Tall, D., & Vinner, S. (1981). Concept image and concept definition in mathematics with particular reference to limits and continuity. *Educational Studies in Mathematics, 12*(2), 151–169.

Tran, V. H., & Vu, T. D. N. N. (2011). *Mathematic textbook grade 11* (5th ed.). Vietnam Education Publishing House.

Trowbridge, D. E., & McDermott, L. C. (1980). Investigation of student understanding of the concept of velocity in one dimension. *American Journal of Physics, 48*, 1020.

Trowbridge, D. E., & McDermott, L. C. (1981). Investigation of student understanding of the concept of acceleration in one dimension. *American Journal of Physics, 49,* 242.

Witzke, I. (2014). Zur Problematik der empirisch-gegenständlichen Analysis des Mathematikunterrichtes. *Der Mathematikunterricht, 60*(2), 19–31.

Witzke, I., & Spies, S. (2016). Domain-specific beliefs of school calculus. *Journal Für Mathematik-Didaktik, 37*(S1), 131–161. https://doi.org/10.1007/s13138-016-0106-4.

Zandieh, M. (2000). A theoretical framework for analyzing student understanding of the concept of derivative. In E. Dubinsky, A. Schoenfeld, & J. J. Kaput (Eds.), *Research in collegiate mathematics education IV* (pp. 103–127). American Mathematical Society.

Lesson Plan: Differential Calculus Through Applications

18

Chu Cam Tho

Lesson Title: The application of (finite) integrals

Abstract: In these two lessons, calculus is applied to authentic physics problems. In the first lesson, these problems involve motion sequences, starting with a simple vertical throw and ending with superimposed motions in one and two spatial directions. In the second lesson, physical force for a non-trivial case is worked out in terms of a long rope that is pulled upwards over a cliff. With the help of further application tasks, the knowledge acquired by students is consolidated. Here, too, finite integrals are necessary for the calculation.

Type of school / Grade	High School / Grade 11
Prerequisites	– definitions of integral and antiderivative – knowledge of the laws of accelerated motion (displacement-time law, velocity, acceleration) and work – familiarity with GeoGebra
Number of periods	2

C. C. Tho (✉)
The Viet Nam Institute of Educational Sciences (VNIES), Research Division On Educational Assessment (RDEA), Hanoi, Vietnam

© The Author(s), under exclusive license to Springer Fachmedien Wiesbaden GmbH, part of Springer Nature 2022
F. Dilling and S. F. Kraus (eds.), *Comparison of Mathematics and Physics Education II*, MINTUS – Beiträge zur mathematisch-naturwissenschaftlichen Bildung, https://doi.org/10.1007/978-3-658-36415-1_18

Objectives	Mathematics: – Students can calculate integrals in simple cases – Students can use integrals to calculate the area of 2D and 3D shapes in simple cases – Students can apply integrals to solve some real-life problems Physics: – Students can interpret the formula for average speed to define speed in one direction – Students can apply the formula to calculate speed and velocity – Students can calculate work in simple cases

18.1 Methodical Commentary

This lesson plan uses the inquiry-based learning approach to connect integral calculus to key concepts in kinematics. This is linked to real-world problems in the form of images of trajectories to objects that exhibit a quadratic relationship. In the second part of the sequence, physical problems concerning Hooke's law and physical work are used as authentic tasks.

From a methodological point of view, we aim for a high degree of student activity, realized by the think-pair-share (TPS) and tablecloth[1] methods as well as individual and group work phases. Furthermore, inquiry-based learning is used in connection with the K-W-L strategy (what I know; what I want to know; what I learned).

Some technical limitations may occur when using the GeoGebra software in Phase 1 of the lesson plan due to an insufficient number of devices or a lack of ICT skills. Special attention should be paid to these aspects.

For the motivating introductory question in the second part of the sequence, the Riemann sum (Fig. 18.1) is used to calculate the physical work required to pull a long rope up over a cliff. For support, it could be useful to provide an additional GeoGebra model to clarify the relationships. It is also likely that misconceptions will occur regarding the variability of force over time.

---

[1] In the tablecloth method, students write down the products of their learning on a piece of A3 paper divided into two sections. The outer area shows individual work, and the central area shows the results of group discussion.

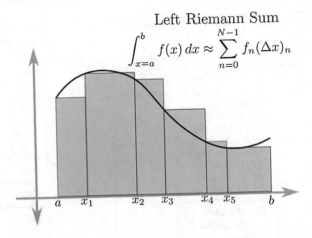

Fig. 18.1 The Riemann sum serves the specification of the description of the area under a function graph

18.2   Lesson Plan—Part 1

No./ Tim-ing	Stage	Learning activities	Interaction form	Materials/ Resources	Methodological comments
# 1 15 min	Motivation	– Students observe pictures of the trajectories of objects and predict which function is suitable for each trajectory – Students determine quadratic functions based on graphs using GeoGebra software – The teacher explains the relation between distance, speed, and acceleration over time based on derivatives	– Teacher presenta-tion – Group work	– Pictures of the trajectories of objects which have the form of a quadratic function – Computer/ phone with GeoGebra pre-installed	– Inquiry-based learning – Students can investigate the problem based on the previous lesson's homework

No./ Timing	Stage	Learning activities	Interaction form	Materials/ Resources	Methodological comments
# 2 10 min	– Introducing new knowledge	– Teacher poses a question: How can we calculate the distance traveled by a moving object or the time it takes a falling object to touch the ground if we know the function of velocity and acceleration over time? – Students discuss based on their knowledge of antiderivatives and definite integrals	– Group work – Teacher presentation		Think-pair-share Think: Teacher begins by asking a specific question: – How does gravity affect a falling object? – Do we always take acceleration a to be equal to g? – What is the relationship between distance, velocity, and acceleration over time, based on antiderivatives and definite integrals? Pair: Each student is paired with another student or a small group Share: Students share their thinking with their partners. The teacher expands the "share" into a whole-class discussion
# 3 10 min	Practice	Students work on the following exercise: A ball is thrown upward with an initial velocity of 64 ft/s from an initial height of 80 feet (a) Find the function giving the height of the ball as a function of time (b) When does the ball hit the ground?	Individual work	Pre-printed learning worksheet	– Students should sketch the model with the x-axis oriented vertically, with the origin at the ground and the upward direction being positive

No./ Timing	Stage	Learning activities	Interaction form	Materials/ Resources	Methodological comments
# 4 10 min	Application	Students complete the exercises on Worksheet 1	Group work	Pre-printed learning work-sheet	– Students may find it difficult to do Exercise 3b – The teacher should help students distinguish between the formula to calculate distance $\int_a^b |v(t)|dt$ and that for displacement $\int_a^b v(t)dt$ of particles over time

Worksheet 1

Use $a(t) = -10$ m/s as the acceleration due to gravity (neglect air resistance) for ex. 1 and ex. 2.

Exercise 1: A ball is thrown vertically upward from a height of 6 m with an initial velocity of 60 m/s. How high will the ball go?

Exercise 2: A balloon, rising vertically with a velocity of 16 m/s, releases a sandbag at the instant it reaches 64 m above the ground.

(a) How many seconds after release will the bag strike the ground?
(b) At what velocity will it hit the ground?

Exercise 3: A particle is moving along a line so that its velocity is $v(t) = t^3 - 10t^2 + 29t - 20$ m/s at time t.

(a) What is the displacement of the particle in the time interval $1 \le t \le 5$?
(b) What is the total distance traveled by the particle in the time interval $1 \le t \le 5$?

Solutions for practice activity

(a) Let $t = 0$ represent the initial time. The two given initial conditions can be written as follows: $s'(t) = 80, s''(t) = 64$, where the initial height is 80 ft and the initial velocity is 64 ft/s.

Using -32 ft/s as the acceleration due to gravity, we can write:

$$s''(t) = 32, s'(t) = \int s''(t)dt = \int -32t\,dt = -32t + C_1$$

Using the initial velocity, we obtain $s'(0) = 64 = C_1$ which implies that $C_1 = 64$. Next, by integrating $s'(t)$, we obtain:

$$s(t) = \int s'(t)dt = \int (-32t + 64)dt = -16t^2 + 64t + C_2$$

Using the initial height, we obtain $s(0) = 80 = C_2$ which implies that $C_2 = 80$. So, the position function is $s(t) = -16t^2 + 64t + 80$

(b) Using the position function found in part (a), we can find the time at which the ball hits the ground by solving the equation for $s(t) = 0$.

$$s(t) = -16t^2 + 64t + 80 = 0 \Rightarrow -16(t + 1)(t - 5) = 0 \Rightarrow t = -1, t = 5$$

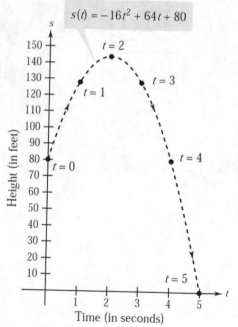

Height of a ball at time t

Because the answer must be positive, we can conclude that the ball hits the ground 5 s after it is thrown.

Solutions for Worksheet 1

Exercise 3: (a) By definition, we know that the displacement is

$$\int_1^5 v(t)dt = \int_1^5 \left(t^3 - 10t^2 + 29t - 20\right)dt = \left[\frac{t^4}{4} - \frac{10}{3}t^3 + \frac{29}{2}t^2 - 20t\right]_1^5$$

$$= \frac{25}{12} - \left(-\frac{103}{12}\right) = \frac{32}{3}$$

So, the particle moves $\frac{32}{3}$ meters to the right.

(b) To find the total distance traveled, we can calculate $\int_1^5 |v(t)|dx$. Using Fig. 18.1 and the fact that the expression t $v(t)$ can be factored as $(t-1)(t-4)(t-5)$, we can determine that $v(t) \geq 0$ at $1 > t > 4$ and $v(t) \leq 0$ at $4 > t > 5$. So, the total distance traveled is:

$$\int_1^5 |v(t)|dt = \int_1^4 v(t)dt - \int_4^5 v(t)dt = \int_1^4 \left(t^3 - 10t^2 + 29t - 20\right)dt$$

$$- \int_4^5 \left(t^3 - 10t^2 + 29t - 20\right)dt$$

$$= \frac{45}{4} - \left(-\frac{7}{12}\right) = \frac{71}{6} \text{ meters}$$

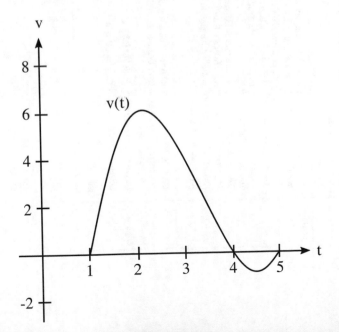

18.3 Lesson Plan—Part 2

No./Time	Stage	Learning activities	Form of Interaction	Materials/Resources	Methodological comments
# 1 10 min	Motivation	– Students state the formula for calculating the work W required to move an object horizontally, based on the impact force F (N) and distance traveled d (m) – The teacher demonstrates how the formula is calculated when force F fluctuates as the result of a function (with one variable) – The teacher gives a specific example: A 60 m climbing rope is hanging over the side of a tall cliff. How much work is performed in pulling the rope to the top, when the rope has a mass of 66 g/m?	– Teacher presentation – Group work	PowerPoint slideshow	– Inquiry-based learning based on the K-W-L strategy: – What do students know about the formula for work as the product of force and distance? – What do students want to know about how the formula changes when the force varies? – What have students learned after using the Riemann sum to calculate the change in work over time? (See Fig. 18.1.)
# 2 10 min	Introducing new knowledge	– Students model the problem by sketching a picture and writing the expression to calculate the work of pulling the rope – Students determine the formula for the general calculation of work $W = \sum_{i=1}^{n} F(x_i)\Delta x_i$. When $n \to \infty$ then $W = \int_a^b F(x)dx$	– Group work – Teacher asks, students answer	Computer/phone with GeoGebra pre-installed	– The teacher may prepare the GeoGebra model of the Riemann sum of the change in work as a function – Limitations: misconception about the consistency of force over time; the difficulty of choosing the variable x and representing the force function

No./Time	Stage	Learning activities	Form of Interaction	Materials/Resources	Methodological comments
# 3 10 min	Practice	Students apply the formula to solve the problem from the beginning of the lesson Let x be the amount of rope pulled in. The amount of rope still hanging is $60 - x$. This length of rope has a mass of $66\,g/m$, or $0.066\,kg/m$. The mass of the remaining rope is $0.066(60 - x)\,kg$. Multiplying this mass by the acceleration of gravity, $9.8\,m/s^2$ gives the variable force function: $F(x) = (9.8)(0.066)(60 - x) = 0.6468(60 - x)$. Thus, the total work performed in pulling up the rope is $$W = \int_{0}^{60} 0.6468(60 - x)dx = 1164.24\,J.$$	Individual work	Pre-printed learning worksheet	Think-pair-share Think: Teachers begin by asking specific questions: – How can we create a force function $F(x)$ for the interval $[0.60]$? – How do we decide what x measures: the length of rope still hanging or the amount of rope pulled in? Pair: Each student is paired with another student or a small group Share: Students share their thinking with their partners. The teacher expands the "share" into a whole-class discussion
# 4 15 min	Application	Students complete Worksheet 2	Group work	Pre-printed learning worksheet	– Tablecloth strategy – Students may find it difficult to remember Hooke's law and its applications – For Exercise 4, students should orient the x-axis vertically, with the origin at the top of the tank and the upward direction being positive

Worksheet 2

Exercise 1: (a) Find the work done on a spring when you compress it from its natural length of 1 m to a length of 0.75 m if the spring constant is $k = 16\,\text{N/m}$.

(b) How much additional work is done if you compress the spring a further 30 cm?

Exercise 2: A force of 1200 N compresses a spring from its natural length of 18–16 cm. How much work is done if it is compressed from 16 cm to 14 cm?

Exercise 3: A leaky bucket weighing 5 N is lifted 20 m into the air at a constant speed. The rope weighs 0.08 N/m. The bucket starts with 2 N of water and leaks at a constant rate. It finishes draining just as it reaches the top. How much work is done:

(a) lifting the water alone?
(b) lifting the water and bucket together?
(c) lifting the water, bucket, and rope?

Exercise 4: Assume a cylindrical tank with a radius of 4 m and a height of 10 m is filled to the 8 m level. How much work does it take to pump all the water over the top edge of the tank, given that the weight density of water is $9800\,\text{N/m}^3$?

Solutions for Worksheet 2

Exercise 1: (a) The force when the spring is extended (or compressed) by x units is given by:

$$F = 16x$$

We start compressing the spring at its natural length (0 m) and finish at 0.25 m from its natural length, so the lower limit of the integral is 0 and the upper limit is 0.25. Therefore:

$$\text{Work} = \int_0^{0.25} 16x\,\mathrm{d}x = \left[8x^2\right]_0^{0.25} = 0.5\,\text{N}\cdot\text{m}$$

(b) This time, we start pushing the spring at 0.25 m from its natural length and finish at 0.55 m from the natural length, so the lower limit of the integral is 0.25 and the upper limit is 0.55.

$$\text{Work} = \int_{0.25}^{0.55} 16x \, dx = \left[8x^2\right]_{0.25}^{0.55} = 1.92 \text{ N} \cdot \text{m}$$

Exercise 2: We must first work out the spring constant (in cm):
$F = kx$. So, $1200 = k \times (2)$. So, $k = 600$ N/cm. In this case, $F = 600x$.
Now the work done is given by the formula:

$$\text{Work} = \int_{2}^{4} 600x \, dx = \left[300x^2\right]_{2}^{4} = 3600 \text{ N} \cdot \text{cm}$$

Exercise 3: (a) The force required to lift the water is its weight.

When the bucket is x meters off the ground, the expression for the weight of the water can be found by first graphing the scenario:

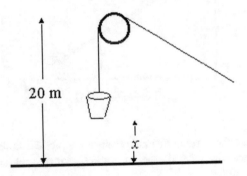

So, using $y = mx + c$, we see that the slope is
$m = -\frac{2}{20} = -\frac{1}{10}$ and the y-intercept is $c = 2$.
So we can write the function for the weight of the water at height x as:

$$F(x) = -\frac{x}{10} + 2$$

Then, $\text{Work} = \int_{a}^{b} F(x) \, dx = \int_{0}^{20} \left(-\frac{x}{10} + 2\right) dx = \left[-\frac{x^2}{20} + 2x\right]_{0}^{20} = -20 + 40$

$$= 20 \text{ N} \cdot \text{m (20 Joule)}$$

(b) For the bucket, $W = Fd = 5 \cdot 20 = 100 \text{ N} \cdot \text{m}$

So, the total work $= 20 + 100 = 120$ N $\cdot$ m

(c) The weight of the rope at height x is:

$$F(x) = 0.08(-x + 20)$$

The work done on the rope is:

$$\int_0^{20} 0.08(-x + 20)dx = 0.08\left[-\frac{x^2}{2} + 20x\right]_0^{20} = -16 + 32 = 16 \text{ N} \cdot \text{m}$$

So the total work done on the water, bucket, and rope is:

$$W = (20 + 100 + 16) \text{ N} \cdot \text{m} = 136 \text{ N} \cdot \text{m or } 136 \text{ J}.$$

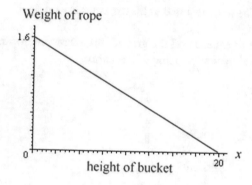

Exercise 4:

We let x represent the vertical distance from the top of the tank. That is, we orient the x-axis vertically, with the origin at the top of the tank and the downward direction being positive.

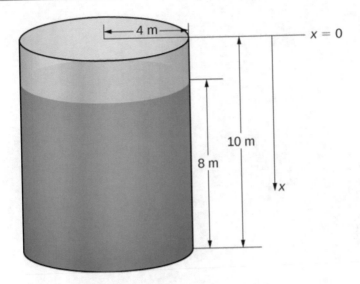

Using this coordinate system, the water extends from $x = 2$ to $x = 10$. In pumping problems, the force required to lift the water to the top of the tank is the force required to overcome gravity, so it is equal to the weight of the water. Given that the weight density of water is $9800 \, \text{N/m}^3$, we can determine:

$$V = \pi 4^2 x = 16\pi x$$

Then, the force needed to lift the water is:

$$F = 16\pi x \cdot 9800 = 156,800\pi x$$

The work done to empty the tank is given by:

$$W = \int_{2}^{10} 156,800\pi x \, dx = 7,526,400\pi \, (\text{J})$$

Therefore, the work required to empty the tank is approximately $23,650,000 \, \text{J}$.

Lesson Plan: Capacitor Charge and Discharge Process. Capacitor Energy

19

Tran Ngoc Chat

Lesson Title: Capacitor charge and discharge process
Abstract: In this lesson, students will learn about the change of voltage on a capacitor over time during the processes of charging and discharging. By applying their mathe-matical knowledge of derivatives, integrals, and some mathematical features of exponential functions, students will determine the rule for the change of voltage over time and the expression used to calculate the energy of the capacitor. The lesson plan also uses several experiments to test students' predictions and aid in the discovery of new rules.

Type of school / Grade	High School / Grade 11
Prerequisites	Mathematics: – Exponential and logarithmic functions – Properties of derivatives and integrals Physics: – Relationship between voltage U, charge q, and capacitance C of the capacitor – Definition of current – Formulas to calculate the work of electric current
Number of periods	2

T. N. Chat (✉)
Hanoi National University of Education, Faculty of Physics, Hanoi, Vietnam
e-mail: chattn@hnue.edu.vn

© The Author(s), under exclusive license to Springer Fachmedien Wiesbaden GmbH, part of Springer Nature 2022
F. Dilling and S. F. Kraus (eds.), *Comparison of Mathematics and Physics Education II*, MINTUS – Beiträge zur mathematisch-naturwissenschaftlichen Bildung, https://doi.org/10.1007/978-3-658-36415-1_19

257

Objectives	Mathematics:
	Students can:
	– Draw the graph of an exponential function
	– Discover that current is the derivative of charge with respect to time
	– Propose a graph representation plan to facilitate finding the function representing the relationship of U and t
	– Identify the graph representing the measurement results of U-t that is similar to the graph of an exponential function
	– Transform the expressions to find an equation representing the relationship between U and t
	– Master the transformations of variables to con-vert complex integral expressions into fundamental functions
	– Generalize a function that represents the relationship between q and t
	– Find the meaning of the time constant
	– Propose a solution to calculate the integral from the way of determining the capacitor's energy
	– Generalize an expression to determine the energy of a capacitor
	Physics
	Students can:
	– Apply knowledge of electrostatics to predict the evolution of charge of a capacitor when charging and discharging
	– Qualitatively argue why the time gap t of the voltage drop is higher when either resistance or capacitance is higher
	– Perceive the relation between voltage and the time gap when the capacitor discharges
	– Propose a theoretical solution to find the relationship between U and t
	– Propose an experimental solution to find the relationship between U and t
	– Explain how to determine the energy of a capacitor
	– Assemble simple electrical elements according to a circuit diagram
	– Adjust the parameters of the electrical elements to measure the appropriate values
	– Apply conclusions from the relationship between U and t to solve some practical problems

19.1   Methodical Commentary

This lesson can be taught along the path of knowledge discovery. To motivate students to seek knowledge, in the first stage a problem situation will be created based on students' misconceptions. The next stage will ask students to propose and implement a plan. Combining physics knowledge acquired in previous lessons with mathematical knowledge about integrals, differentials, and the characteristics of exponential and logarithmic functions, students will be able to solve problems. To make the content meaningful for students, the teacher will have the students apply the RC circuit laws to make a timer.

Problem discovery phase: Students know the structure of a capacitor from the previous lesson. They may have the misconception that, when a capacitor is discharged, the charge is neutralized immediately because the charge at the two poles of the capacitor are opposite in sign. However, in the experiment, when discharging the capacitor through the light bulb, they will see that the current does not decrease immediately. The experiment also shows that the change of voltage U depends on the magnitude of the resistance R and that of the capacitance C of the capacitor. The question is: What law does the voltage obey with regard to time?

The solution proposal and solution implementation phase will comprise two approaches. In the first, students will conduct an experiment to measure voltage U and time t. They will utilize the graphing software on the computer to find a suitable function. The second approach is to apply the physics knowledge students have learned:

$$I = \frac{U}{R}$$

$$q = C \cdot U$$

and treat current as the derivative of the function q over time t:

$$I = \frac{dq}{dt}$$

On this basis, an equation showing the relationship between U and t can be drawn. The relationship can be found by solving this equation, using the mathematical rule for finding the primitive function.

The theoretical approach is consistent with the experimental approach, thereby helping students see the close connection between mathematics and physics. The approach that uses mathematical knowledge about derivatives and integrals will help students discover the rule for the change in voltage in the process of charging the capacitor and the expression to determine the capacitor energy. Using this approach, students will acquire a wide range of new physics knowledge.

The knowledge expansion phase can be approached through the deepening understanding of the time constant quantity $T = R \cdot C$ in the charge or discharge circuit of the capacitor. Students should notice that the time gap is proportional to the magnitude of the resistor. On that basis, students will design a timer circuit that can be used to adjust the time that a light remains lit by adjusting the value of the resistor using the rheostat.

19.2 Lesson Plan

Lesson Plan

Nr./Time	Stage	Learning activities	Interaction form	Materials/ Resources	Methodological comments
#1 5 min	Ensure background knowledge	The teacher assigns students the tasks: State the relationship between voltage U, charge q, and capacitance C of a capacitor Write an equation to represent electric current energy	Teacher present Whole class work		Make sure students have a clear understanding of the fundamentals: $C = \frac{U}{q}$ Equation to represent electric current energy: $E = U \cdot q = U \cdot I \cdot t = I^2 \cdot R \cdot t$
#2 15 min	Detect the problem	The teacher asks the question: "When the capacitor discharges, will it lose power suddenly or will it decrease smoothly over time, and why?" Students answer. The teacher introduces and conducts Experiment 1. Students observe the experiment to verify their predictions The teacher divides the class into four groups Students conduct Experiment 2. The teacher reads the experimental results of the first pair of U and t values, and students predict the next U and t values. Then students conduct the experiment, read the results and compare them to their predictions. The teacher asks students to comment Students predict, record, and compare experimental results	Teacher present Whole class work Group work	Experiment 1 Worksheet 1 Experiment 2	Experiment 1 helps students see the fact that when a capacitor discharges, the charge on the capacitor is not neutralized immediately but decreases smoothly over time In addition, help students see the preliminary relations: The voltage on the capacitor decreases more slowly when the capacitance value or the resistance value is greater The results of this experiment also create favorable conditions for students to make hypotheses in the next section From Experiment 1, students will see that U decreases with t, which easily leads to the false hypothesis that U is inversely proportional to t Experimental results will disprove this hypothesis

Nr./Time	Stage	Learning activities	Interaction form	Materials/ Resources	Methodological comments
		The teacher asks students to explain the difference between predicted and experimental results Students explain and find the problem The teacher states the problem formally			Contradictions arise so students will be motivated to determine the specific relationship between U and t when the capacitor discharges The teacher can suggest a preliminary way for students so that students also have the opportunity to propose research solutions in the following stages: According to Ohm's law and by definition of current, $$I = \frac{U}{R} = \left|\frac{\Delta q}{\Delta t}\right| = \left|\frac{\Delta C \cdot U}{\Delta t}\right| = C \cdot \left|\frac{\Delta U}{\Delta t}\right|$$ That is: $\left|\frac{\Delta U}{\Delta t}\right| \sim U$. Since U decreases with time, the rate of voltage drop $\left|\frac{\Delta U}{\Delta t}\right|$ also decreases over time (i.e., it is not constant). Therefore, U does not decrease inversely with time In addition, in Experiment 1, it was shown that the process of decreasing voltage U over time t is regular. The rate of reduction depends on the magnitude of C and R. However, the decrease in U is not proportional to t. So what is the specific reduction law of U with respect to t? Because of this situation, students will be motivated to deeply understand the relationship between U and t As a result, teachers can formally state the problem as: "What is the specific mathematical relation of the voltage on the capacitor over time when the capacitor is discharging?"

Nr./Time	Stage	Learning activities	Interaction form	Materials/ Resources	Methodological comments
#3 10 min	Proposing a solution to the problem	The teacher asks prompting questions so that students can suggest solutions Students propose experimental solutions: From the measured data, use graphing software to find the function representing the relationship between U and t. Change the values of C and R to find the general rule Students propose theoretical solutions: From equations learned representing relationships, apply mathematical knowledge to find a function representing the relationship between U and t The teacher assigns each group to implement both solutions. Compare the results of the two solutions to draw conclusions	Teacher present/ Whole-class work		For the experimental solution: the teacher can suggest that when it is necessary to find the specific relationship of two measured quantities, software to find the FIT function can be used For the theoretical solution: from the previous stage that suggests the expressions representing the relationship of related quantities, students can now suggest a way to solve the problem Teachers also need to remind students of some related math knowledge, including: Definition of current as the derivative of charge with respect to time: $$I = \lim_{\Delta t \to 0} \frac{\Delta q}{\Delta t} = \frac{dq}{dt}$$ $$\ln a = b \Leftrightarrow a = e^b$$ $$\ln' x = \frac{1}{x} \Rightarrow \frac{dx}{x} = d(\ln x)$$

Nr./Time	Stage	Learning activities	Interaction form	Materials/Resources	Methodological comments
#4 25 min	Implement the solution	Experimental solution: Students conduct experiments and collect experimental data Students process experimental results with the help of computers and draw the relationship between U and t Theoretical solution: Students transform formulas and apply mathematical knowledge to find the relationship between U and t Compare the relationship between the experimental results and the results of the mathematical transformation. Draw conclusions about the function representing the relationship between U and t	Group work	Experiment 3 Software helps the students draw the graph and find the FIT function (e.g., Excel, Origin, Coach, etc.) Worksheet 2	Note that the experimental solution only helps to find the form of the dependence function between U and t: an exponential function with parameters that are specific numbers, not an expression representing the relationship of the variables of physical quantities The theoretical solution can be implemented as follows: $I = \frac{U_{AB}}{R} = \frac{dq_B}{dt} = -\frac{dq_A}{dt} = -\frac{d(C \cdot U_{AB})}{dt}$ $\Rightarrow \frac{dU_{AB}}{U_{AB}} = -\frac{dt}{R \cdot C}$ $\Rightarrow d(\ln U_{AB}) = d\left(-\frac{t}{R \cdot C}\right)$ $\Rightarrow \ln U_{AB} = -\frac{t}{R \cdot C} + const$ $\Rightarrow U_{AB} = U_o \cdot e^{-\frac{t}{R \cdot C}}$ In which, U_o is the value of the potential difference at the time the capacitor starts to discharge (i.e., $t = 0$ s) The theoretical and experimental results are compared. From there, students draw a general mathematical relationship between U and t when the capacitor discharges

Nr./Time	Stage	Learning activities	Interaction form	Materials/ Resources	Methodological comments
#5 5 min	Problem development	The teacher develops the problem, asking the students to: * Find the physical meaning of the function expression * Determine the energy expression of the capacitor * Find the relationship of U and t for the capacitor charging case. Students apply mathematical knowledge to find the answers	Group work	Worksheet 3	Students apply mathematical knowledge to perform tasks (details on Worksheet 3)
#6 5 min	Conclusion	The teacher shows students the consistency of research results and results in textbooks. The teacher determines the conclusion of the lesson and gives definitions of quantities. Students state and record important conclusions of the lesson	Teacher present / Whole class work		The voltage across the capacitor during charging and discharging varies with time according to the law of an exponential function. In case of discharge: $$U(t) = U_o \cdot e^{-\frac{t}{RC}}$$ In the case of an electric charge: $$U(t) = U_o \cdot \left(1 - e^{-\frac{t}{RC}}\right)$$ After a fixed amount of time, the rate of change of voltage across the capacitor is determined. The magnitude of that ratio depends on the product of R and C. Therefore, one calls $T = R \cdot C$ a time constant. The energy of the capacitor is: $$E = \frac{1}{2}\frac{q^2}{C} = \frac{1}{2}C \cdot U_o^2$$

Nr./Time	Stage	Learning activities	Interaction form	Materials/ Resources	Methodological comments
#7 10 min	Practice	The teacher corrects problem #1 on the worksheet Students practice solving some basic problems on Worksheet 4	Teacher present/ Group work	Worksheet 4	The teacher conducts a sample solution of an illustrated problem to help students familiarize themselves with new concepts and the type of problem
#8 15 min	Extension	The teacher asks students to design a timer device based on the rule for the changing value of the voltage on the capacitor when charging or discharging	Teacher present/ Whole class work	Experi- ment 4	Based on the law of variation of voltage, guide students to design and test a timer circuit. For example, the timer turns an electrical device on or off In this section, transistors will be introduced. Students have not studied transistors, so the teacher needs to introduce their effects and operating conditions in a simple way as an introduction to Experiment 4 (see Appendix)

Experiment 1
Experiment 1: The capacitor discharges electricity through the light bulb

Experiment circuit:

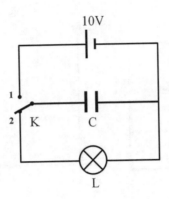

Case 1:
Charge the capacitor by turning the contact K to Position 1. Then discharge the capacitor through the lamp L by turning the contact K to Position 2. Observe the changes in the brightness of the lamp.

Case 2:
Arrange the devices the same as in Case 1, but double the capacitance of the capacitor by pairing 2 capacitors in parallel. Observe the changes in the lamp's brightness and compare it with Case 1.

Case 3:
Arrange the devices the same as in Case 1, but reduce the external circuit resistance by half by pairing two identical bulbs in parallel. Observe the changes in the lamps' brightness and compare it with Case 1.

Experiment results:
Case 1: The brightness of the light bulb decreases over time and then turns off completely. From this, it is clear that the current going through the lamp and the voltage on the capacitor decrease with time.

   Cases 2 and 3: It is observed that in Case 2 the time that the lamp is lit is longer than the time in Case 1 (i.e., the voltage (and hence current) decreases more slowly). In Case 3, we see the opposite process.

Worksheet 1

Experimental Investigation 2: Measuring voltage U over time t when the capacitor discharges.

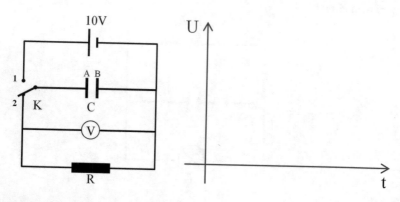

Task 1: Conduct the experiment. Record U_1 and t_1, the values of the first measurement. Guess the time values on the table. Plot your predictions.

Table 1 Prediction of experiment results

No.	0	1	2	3	4	5	6	7	8	9	10
U (V)	10	9	8	7	6	5	4	3	2	1	0.5
t (s)	0										

Task 2: Conduct the experiment to measure the corresponding time values and record them in the table of values. Plot the measured values on the same coordinate plane as your predictions.

Table 2 Measurement results from the experiment

No.	1	2	3	4	5	6	7	8	9	10
U (V)	0.1	0.2	0.4	0.8	1	2	3	4	5	6
I (mA)										

Task 3: Compare the predicted and measured results from the experiment. Can you explain the difference? Give a problem statement.

Worksheet 2

Experimental Investigation 3: Measure voltage U and time t when a capacitor discharges with different capacitance and resistance values.

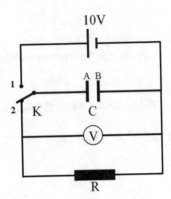

Task1: Conduct the experiment. Record U and t values for the cases where R doubles and C doubles. Enter the measured values into the graphing software to find the corresponding FIT function. Compare the FIT functions obtained between different R, C values. Comment on the result.

Table 3 Prediction of experimental results

No.	0	1	2	3	4	5	6	7	8	9	10
U (V)	10	9	8	7	6	5	4	3	2	1	0.5
t_a (s)	0										
t_b (s)	0										

Task 2: Mathematical transformation to find a function representing the relationship between U and t.

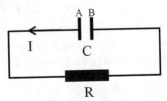

Task 3: From the results of Task 2, infer experimentally measurable consequences for Task 1. Compare the results from experiments and the results inferred from mathematical transformations. Write a comment.

Worksheet 3

Task 1: Present the physical meaning of the function representing the relationship between U and t during the discharge of a capacitor with the initial voltage U_o.

$$U = U_o \cdot e^{-\frac{t}{RC}}$$

Task 2: Describe how to build a mathematical expression to determine the energy of the capacitor.

Task 3: Describe how to find the relationship between U and t for the case of charging the capacitor.

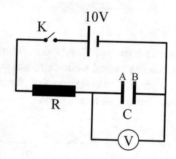

Worksheet 3 (Methodological comments)

Task 1: Present the physical meaning of the function representing the relationship between U and t during the discharge of a capacitor with the initial voltage U_o

$$U = U_o \cdot e^{-\frac{t}{RC}}$$

$$\Rightarrow q = q_o \cdot e^{-\frac{t}{RC}} \Rightarrow I = I_o \cdot e^{-\frac{t}{RC}}$$

The voltage decreases according to the law of an exponential function, so the charge on the capacitor plate and the current decrease according to the same law. In addition, the expression representing the potential difference over time is:

$$\frac{t}{R \cdot C} = \ln\left(\frac{U_o}{U}\right)$$

This expression shows that, for any given time, the ratio of the voltage drop across the capacitor is determined; it does not depend on the initial value of the voltage. This voltage drop ratio depends only on the product of R and C.

Task 2: Describe how to build a mathematical expression to determine the energy of the capacitor.

Method 1 The energy of the capacitor is determined as the discharge current energy during the whole process. This current I passes through the resistor R, since the power of the current is:

$$P = I^2 \cdot R = \frac{U^2}{R} = \frac{U_o^2}{R} \cdot e^{-\frac{2 \cdot t}{R \cdot C}}$$

So, the total energy of the capacitor is:

$$E = \lim_{\Delta t \to 0} \sum_{t=0}^{t=\infty} P \cdot \Delta t = \int_0^\infty P \cdot dt = \int_0^\infty \frac{U_o^2}{R} \cdot e^{-\frac{2 \cdot t}{R \cdot C}} \cdot dt \qquad (19.1)$$

$$\Rightarrow E = \frac{U_o^2}{R} \cdot \left(-\frac{R \cdot C}{2} \right) \int_0^\infty e^{-\frac{2 \cdot t}{R \cdot C}} \cdot d\left(-\frac{2 \cdot t}{R \cdot C} \right)$$

$$\Rightarrow E = -\frac{C \cdot U_o^2}{2} \cdot e^{-\frac{2 \cdot t}{R \cdot C}} \Big|_0^\infty$$

$$\Rightarrow E = \frac{C \cdot U_o^2}{2} \qquad (19.2)$$

When we present Eq. (19.1) and (19.2) on the graph, we see that the capacitor energy is the area S_2 created by the function $P(t)$ which has the same value as the area of the triangle S_1.

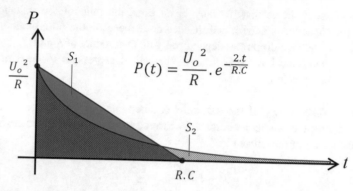

$$P(t) = \frac{U_o^2}{R} \cdot e^{-\frac{2.t}{R.C}}$$

Method 2 The energy of the capacitor is the electric field energy that moves the charges.

$$E = \lim_{\Delta q \to 0} \sum_{q=0}^{q=q_o} U \cdot \Delta q = \int_0^{q_o} U \cdot dq = \int_0^{q_o} \frac{q}{C} \cdot dq$$

$$\Rightarrow E = \frac{q^2}{2 \cdot c} \bigg|_0^{q_o}$$

$$\Rightarrow E = \frac{q_o^2}{2 \cdot C} = \frac{C \cdot U_o^2}{2} \tag{19.3}$$

When expressing Eq. (19.3) on the graph, the capacitor energy is the area of the triangle bounded by the graph $U(q)$.

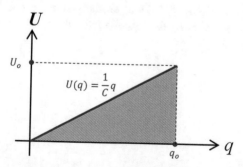

Task 3: Describe how to find the relationship of U and t for the case of charging the capacitor.

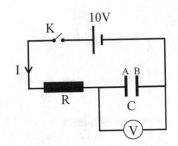

$$I = \frac{U_o - U_{AB}}{R} = \frac{dq_A}{dt} = \frac{d(C \cdot U_{AB})}{dt}$$

$$\Rightarrow \frac{d(U_o - U_{AB})}{U_o - U_{AB}} = -\frac{dt}{R \cdot C}$$

$$\Rightarrow d(\ln (U_o - U_{AB})) = d\left(-\frac{t}{R \cdot C}\right)$$

$$\Rightarrow \ln (U_o - U_{AB}) = -\frac{t}{R \cdot C} + \text{const}$$

$$\Rightarrow U_{AB} = U_o\left(1 - e^{-\frac{t}{R \cdot C}}\right)$$

Worksheet 4

Solve the following problems:

Problem 1: A capacitor with capacitance $C = 1000\,\mu F$ is initially charged to a voltage of 9 V. This capacitor is discharged through a resistor $R = 20\,k$.

(a) Determine the time constant T of the capacitor.
(b) After time T, what is the voltage across the capacitor?
(c) How long does it take for the voltage on the capacitor to halve?

Problem 2: Consider an electrical circuit with the design shown in the figure. Initially, the contact K_1 is switched to Position 1 to charge the capacitor. Then it is switched to Position 2 to discharge the capacitor. At the initial time contact K_2 is in the on position. Know that $C = 2000\,\mu F$, $R_1 = 10\,k$ and $R_2 = 5\,k$.

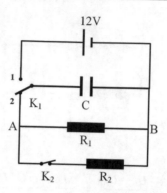

(a) How long after switching contact K_1 to Position 1 does $U_{AB} = 9\,$V?
(b) At the time $U_{AB} = 9\,$V we open K_2. How long does it take for the voltage to reach $U_{AB} = 6\,$V after opening K_2?
(c) When $U_{AB} = 6\,$V we disconnect the key K_1 to finish the discharge of the capacitor. Determine the amount of heat that the capacitor discharge current has provided for the two resistors R_1 and R_2.

Problem 3: Consider an electrical circuit as shown in the figure. Initially, the key K_1 is switched to Position 1 to charge the capacitor C_1. Then it is switched to Position 2 to discharge the capacitor C_1 and, at the same time, charge the capacitor C_2. Consider that $R = 2\,$k, $C_1 = 1000\,\mu F$, and $C_2 = 2000\,\mu F$.

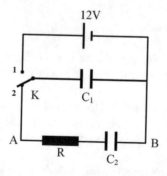

(a) How long after discharge for capacitor C_1 does $U_{AB} = 9\,$V?
(b) Determine the amount of heat that the discharge current of the capacitor C_1 provides to the resistor R.

Experiment 4 Illustration of a timer device
Layout and assembly of components such as an electrical circuit

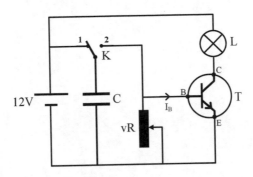

Implementation and results of the experiment:
Adjust the variable resistor (vR) to about 5% of the maximum value. Switch the contact K from Position 1 to Position 2. We see that the lamp (L) is bright. After about 10 s, the light turns off.

Adjust the vR to a higher value and repeat the experiment. We see that the lamp L stays bright for a longer time than it did previously. The greater the vR value, the longer the light stays on.

Thus, by adjusting the magnitude of the variable resistor vR, we can time the lamp to turn off.

Explanation:
Transistor T opens when the voltage U_{BE} has a value greater than 0.6 V, turning L on. When the voltage U_{BE} is less than 0.6 V, T closes, turning L off. The current I_B, which controls the opening and closing of T, is very small, so it does not significantly affect the discharge circuit of capacitor C into the rheostat vR.

When K is in Position 1, the voltage across the capacitor C is loaded to $U_C = 12$ V, $U_{BE} = 0$ V, so T closes as the light turns off.

When we switch contact K_1 to Position 2, then $U_{BE} = U_C$, because initially $U_{BE} = U_C = 12$ V > 0.6 V. T opens, turning L on. However, the capacitor discharges through the resistor vR, so the voltage on the capacitor and U_{BE} gradually decrease. When U_{BE} drops below 0.6 V, T closes and L turns off. The time it takes for U_{BE} to decrease from 12 to 0.6 V depends on the time constant RC. Therefore, by adjusting the magnitude of vR, we can adjust how long the light stays on.

Comparison: Stochastics with a Focus on Probability Theory

20

Gero Stoffels and Sascha Hohmann

Stochastics is a major field in mathematics and highly applicable to physics. Stochastics is usually divided into statistics and probability theory. Statistics is bound to measuring various processes in the empirical sciences and providing methods for the description of datasets, hypothesis building, and testing using inferential statistics. In probability theory, the concept of random processes, their mathematization, and their application to real-world phenomena are the focus. Especially in modern physics, the non-deterministic nature of various physical phenomena (e.g., thermodynamics, quantum theory) shows a need for a deeper understanding of concepts such as probability and chance, alongside their importance for citizens of modern societies. In this article, we focus on probability theory, though we do not underestimate the importance of statistics, especially in the contexts of big data and the need for statistical literacy in the current pandemic.

Therefore, we will give two perspectives, one from mathematics education research and the other from physics education research. We will provide examples of the application of the concepts of probability, give an overview of the (mis-)conceptions identified by each research community, and summarize each perspective's answer to the question, "Why are probability theory and statistics important?" In the final section, the two perspectives will be integrated.

G. Stoffels (✉)
University of Siegen, Mathematics Education, Siegen, Germany
e-mail: stoffels@mathematik.uni-siegen.de

S. Hohmann
IPN—Leibniz Institute for Science and Mathematics Education, Physics Education, Kiel, Germanye-mail: hohmann@leibniz-ipn.de

© The Author(s), under exclusive license to Springer Fachmedien Wiesbaden GmbH, part of Springer Nature 2022
F. Dilling and S. F. Kraus (eds.), *Comparison of Mathematics and Physics Education II*, MINTUS – Beiträge zur mathematisch-naturwissenschaftlichen Bildung, https://doi.org/10.1007/978-3-658-36415-1_20

20.1   Introduction

"6. MATHEMATICAL TREATMENT OF THE AXIOMS of PHYSICS.
The investigations on the foundations of geometry suggest the problem: To treat
in the same manner, by means of axioms, those physical sciences in which mathematics plays an important part; in the first rank are the theory of probabilities and
mechanics."
(Hilbert, 1902, p. 454)

The above quote from Hilbert's (1902) Paris speech about 23 mathematical problems is well known and can be seen as the starting point of the endeavor for the
modern axiomatization of probability theory. The urge to develop such an axiomatization is based, on the desire to systematize mathematics and natural sciences
and, on the particularities of the concepts involved in probability theory, which
seemed "strange" (Czuber, 1898, p. 1). It is also well known that Kolmogoroff
"solved" the problem of axiomatizing probability in his 1933 "Grundbegriffe
der Wahrscheinlichkeitsrechnung" (Foundations of the Theory of Probability) by
skipping a contentual definition of probability. Therefore, even today there are
multiple views on probability (cf. 1.2.2).

Another remarkable aspect of Hilbert's (1902) quote is that he states that probability theory is a physical science. This is reasonable from a nineteenth-century
perspective (cf. 1.3.1) when physics made major progress in thermodynamics
with its subdiscipline of statistical mechanics. Thus, there is also a historical reason to discuss aspects of probability theory from the point of view of mathematics education research and physics education research.

This historical classification is, of course, not the only benefit of such a multi-
perspective view of probability theory and statistics. Both mathematical fields are
applied in physics, and typical activities in physics (e.g., conducting experiments,
gathering and analyzing data) become part of mathematics teaching and learning
in these fields.

In the following sections, both perspectives—one of mathematics education
research and the other of physics education research—will be shown by giving
examples of the application of the concepts of probability and statistics. After this
we will give a detailed description of the contents of probability and statistics as
well as an overview of the (mis-)conceptions identified by each research community. Then, the use of these concepts in school and university will be depicted.
After this, the perspectives will be summarized in terms of how they answer the
question, "Why are probability theory and statistics important?" In the last section, the perspectives will be integrated.

20.2   Mathematics Education Research

The mathematical fields of probability theory and statistics and geometry are the only content areas—at least in Germany—which must be taught from primary school through upper secondary school. In particular, the turn to a more application- and competence-oriented approach to mathematical learning and teaching in response to the results of the first PISA studies (Deutsches P.K., 2013) shows the need for the early implementation of probability and statistics. In the United States, the National Council of Teachers of Mathematics (NCTM, 2000) recommends introducing basic concepts such as likelihood, predictions in simple experiments, and measurement intervals in Grades 3–5. In Germany, the Kultusministerkonferenz (KMK, 2004) gives similar advice, additionally addressing different representations of descriptive statistics.

In lower and upper secondary schools, the topics of probability and statistics have a longer tradition (cf. Batanero et al., 2016). Digital tools and media have provided a new research field of great importance for schools (Biehler & Engel, 2015), as well as the possibility of simulating random processes, using calculus concepts to discuss probability problems, visualizing different probability distributions, and performing numerical methods more often at school.

At university, students of mathematics usually must take at least one course in probability or statistics. Even before students start school, they already might have some ideas about probability and how it can be described. Freudenthal summarizes this pointedly as follows:

> "I already know what a point, a right angle, or a circle is before the first lesson of geometry starts. I simply can't specify them yet. Likewise I already know what probability is before I have defined it."
> (Freudenthal, 1963, S.7, translated by G.S.)

Considering the historical development of probability theory and statistics, as well as today's common treatment of these fields, makes it obvious that they are separate fields. Of course—especially in interferential statistics—the overlap between the two areas are utilized to provide new insights on evaluating certain events as well as testing manifold hypotheses.

Therefore, there are two fields in mathematics education research, which can be loosely summarized as research concerning "statistical literacy" (cf. Ben-Zvi & Garfield, 2004) and "probabilistic thinking" (cf. Chernoff & Sriraman, 2014). For methodological work in physics, statistical literacy may be more relevant. For conceptualizing physical phenomena involving some sort of "randomness" or "indetermination," the probabilistic view seems to be more important. Therefore, the following section will focus on "probabilistic thinking."

20.2.1 A Mathematical (?) Example: Casting a Die

There is a high probability that almost all readers have cast dice (or at least a die). Their experience may come from games like "Ludo" or "Mensch ärger Dich nicht." For illustrating basic concepts of probability theory it is quite useful to explore such games.

In the game "Mensch ärger Dich nicht," it is necessary to cast a common six-sided die, with each player trying to "roll a six" to move their token "out of their house." Each player has three tries before the next player's turn. If someone observes people playing this game, they might hear phrases like,

- "There are only 'ones' on this die,"
- "Casting a 'six' is very unlikely,"
  or maybe, when everybody except one player is already "out of their house,"
- "I will never get out 'of my house.'"

Even from these intuitive statements about probability and chance, as well as the rules of the game, one can easily infer some properties of the random trial in this introductory sequence of the game.

The introductory sequence can be described as a multistage random trial where each stage has the sample space $\Omega = \{1, 2, 3, 4, 5, 6\}$. The symbolic notation for this multistage sample space is $\Omega^n = \Omega \times \Omega \times \ldots \times \Omega = \{1, 2, 3, 4, 5, 6\}^n$. In this example, n is the number of casts needed until the outcome "six" occurs. Based on this discussion of the sample space, it is already possible to declare that the first claim—there are just "ones" on the die—is an exaggeration; if not, it would be good advice to change the die if one is to ever "escape."

The other claims address the probability of certain events. There are different views on probability, which will be discussed in more detail from a mathematics educational perspective in 1.2.2. A common assertion for a case like this example is that the probabilities of the different elementary elements $\{\omega\}, \omega \in \Omega$ with an "ideal" die is given by $P(\{\omega\}) = \frac{1}{6}$ for the single stage random trial. The term "ideal" is an indicator of the "classical interpretation" of probabilities (cf. 1.2.2). Examining this symbolic representation of the probability shows that the probability is not a function with the domain Ω. Instead, the domain of the probability measure P is $\mathcal{P}(\Omega)$. The reason for this—at first glance—complicated definition is based on the desire to assign probabilities not only to elementary events, like "one" and "two," but also to combined events—for example, "not casting a six" = $\{1,2,3,4,5\}$.

So, the probability of one stage of the random trial can be given by the following probability space $(\Omega, \mathcal{P}(\Omega), P)$, using the method of "The Construction of Fields of Probability" (Kolmogorov, 1956, p. 3), with

$$\Omega = \{1, 2, 3, 4, 5, 6\}; \, P(\{\omega\}) = \frac{1}{6}, \omega \in \Omega$$

The second claim of the discussion can now be easily answered. Referring to our model, the probability of casting a "six" is given by $P(\{6\}) = \frac{1}{6}$, which is equal to the probabilities of the other elementary events. In this scenario, the combined event "not casting a six" might be more relevant for the player. Using the model above, the probability for this event is given by $P(\{1, 2, 3, 4, 5\}) = P(\Omega \setminus \{6\}) = \frac{5}{6}$, which is of course five times greater than its complement.

The last claim is still unanswered. Perhaps in probability theory, an answer can be found. First, it is necessary to extend our model. Our sample space is already given by Ω^n. The probability of an elementary event $\omega \in \Omega^n$ is given by

$$P(\{\omega\}) = \left(\frac{1}{6}\right)^n$$

With this result, it is no easier to calculate the probability of the event "I will never get out of my house." We need some knowledge about conditional probabilities. The condition for "never getting out of one's house" is that every previous cast of the die is "not a six." Assuming the die casts are independent, we can then deduce that the probability of the event "I will never get out of my house" is given by

$$\lim_{n \to \infty} P(\{(x_1, \ldots, x_n) : x_i \in \{1, 2, 3, 4, 5\}\}) = \lim_{n \to \infty} \left(\frac{5}{6}\right)^n = 0$$

This can be then interpreted to mean that it is very unlikely that the player will "never get out of their house."

Obviously there is more mathematical knowledge of probability needed to deal with these claims, as they were made explicit in this example. However, the insight gained from this example should be that even starting with simple games, in probability theory, a lot of mathematics is involved and some of the results might be counterintuitive. To emphasize this and the need for mathematics education research, a final point can be made: It is of course possible that the last player will never "get out of his house," but the probability of this is vanishingly small.

20.2.2 Concepts of Probability

As the example in the previous section shows, there is great potential for mathematics education research, which has a "fairly long history" (Batanero, Chernoff et al., 2016) of using probability theory. The example addresses the modelling of real-life phenomena, examines (in-)finite processes, and offers semiotic perspectives of mathematical objects.

Especially, the concepts of independence and conditional probability give insights into everyday contexts. An example may be the high rates of false positive COVID-19 tests during the pandemic. This topic was widely discussed in the media after the introduction of rapid tests. It also gives a further argument for connecting statistics and probability.

During the past year, many papers have been written addressing basic probabilistic concepts in this context (Seifried, et al., 2021; Ramdas, et al., 2020). Before starting to think about these concepts, it should be stated that every test can have a false positive result (meaning the test gives the result "infected" and the person tested is not infected) and a false negative result (meaning the test gives the result "non-infected" and the person tested is infected). A very important aspect of interpreting the probability of these events is their dependence on the prevalence of the disease in the tested population—a statistical measurement. In the case of the COVID-19 pandemic, there are still no reliable numbers for this data. So, three examples comparing different degrees of prevalence and levels of test sensitivity are given. For a better understanding of the connections of the events, a tree diagram is given in Fig. 20.1. To compare the expected outcomes of the tests, two tables (Tables 20.1 and 20.2) are given. Our examples are based on the estimates of the "Seroprävalenz COVID-19 Düsseldorf: SERODUS I & II: Feldbericht und vorläufiger Ergebnisbericht" (Backhaus et al. 2021). The prevalence of COVID-19 in Germany is estimated to be 3.3–9.6%. For each level of prevalence, the test outcomes of 10,000 participants are determined. Schlenger (2020) estimates the "effective sensitivity" of the test to be 71–98%, with a specificity of 95%. The dependence on prevalence can be seen by comparing Table 20.1 with Fig. 20.1 and the dependence of test sensitivity can be seen by comparing Table 20.2 with Fig. 20.1.

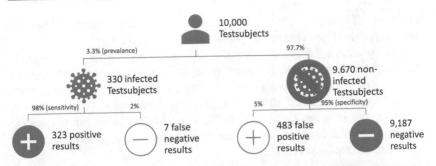

Fig. 20.1   Tree diagram based on prevalence: 3,3%, test's sensitivity: 98%, test's specificity: 95%

Table 20.1   Based on prevalence: 9.6%, test sensitivity: 98%, test specificity: 95%

	Infected person	Non-infected person	Σ
Positive result	941	452	1,393
Negative result	19	8,588	8,607
Σ	960	9,040	10,000

Table 20.2   Based on prevalence: 3.3%, test sensitivity: 71%, test specificity: 95%

	Infected person	Non-infected person	Σ
Positive result	234	483	717
Negative result	96	9,187	9,283
Σ	330	9,670	10,000

This example also demonstrates the importance of using different representations in probability theory and statistics (cf. Trevena et al., 2013).

Much research has been done on the interpretation and use of different conceptions of probability, which are interwoven with philosophical and physical ideas in interesting ways. In the literature, there are seven different common interpretations (intuitive, classical, frequentist, subjective, logical, propensity/objective, and formal–abstract) of probability addressed (cf. Batanero et al., 2016; Biehler & Engel, 2015). In the following section, an outline of these different interpretations is given.

Intuitive ideas about chance emerged very early in history, especially in games. Also, children develop intuitive ideas about the probabilities of winning and losing. The term intuitive in this sense means holding a "degree of belief in the occurrence of random events" (Batanero et al., 2016, p. 3).

The classical definition of probability, which was refined and popularized by Laplace in 1814 in his *Philosophical Essay on Probability*, says that probability is simply a fraction: the number of desired outcomes of a random experiment divided by the number of all possible outcomes. Problems with this view were already apparent when it was published because Laplace states that the assumption of equiprobable events is based on the not (yet) known deterministic background of the random events.

On the contrary, the basic idea of the frequentist interpretation of probability is that the relative frequency h_n should approximate the theoretical probabilities for a large number of trials. This idea is visible in Bernoullis's (1713) "Ars Conjectandi." Von Mises (1928/1952) defines probability as the hypothetical number towards which the relative frequency tends when a random experiment is repeated infinitely many times, based on his theory of collectives. This interpretation is often used in physics and fits perfectly to empirical measuring processes of physical quantities. However, the frequentist approach seems to be adequate for learning probability theory. Batanero et al., (2016, p. 4) offer the assessment:

> "Consequently, it is important to make clear to students the difference between a theoretical model of probability and the frequency data from reality used to create a model of probability. Sometimes this difference is not made explicit in the classroom and may confuse students who need to use abstract knowledge about probability to solve concrete problems from real life."
> (Batanero et al., 2016, p. 4)

This is also a well-known problem in physics education research when it comes to the teaching of physical models and their application to real phenomena.

Popper (1959) introduced the idea of propensity as a measure for the tendency of a random experiment to give a certain outcome. He insists that many frequentists seem to have this idea in mind when they talk about probability. For example, von Mises (1930) states that people have designed dice in such a way that they hold nearly equiprobable outcomes after learning that this property is not dependent on the material used or magic spells. Although propensity seems to be a property of dice, it is not clear if and how it is possible to derive a probability for a single event. For von Mises, such a question would not be fitting to probability theory in general, because he beliefs, that you can only work with probabilities after you have identified a proper collective, which is—by definition—a series of trials.

The interpretations of probability that use logical reasoning were developed by Keynes (1921) and Carnap (1950), who wanted to conserve the a priori nature of the classical interpretation. They propose to logically infer probabilities using evidence and deciding on this basis which hypotheses might be held. Because of the need for formal logical deductions, this approach is not often used in mathematics education although it is quite interesting from a mathematical standpoint.

An approach more often used in schools and universities is the Bayesian approach, also called subjective interpretation. Its foundation is the Bayes Theorem, which can be used to revise probabilities assigned in a random experiment in the light of new data for the outcomes. Especially in fields such as AI, economics, and neuronal networks, the Bayesian approach is important for application (cf. Lampinen & Vehtari, 2001). The interesting thing about the Bayesian approach is that it transfers (guessed) a priori probability to a posteriori probability based on new data, which leads to a non-objective account of probabilities (cf. Batanero et al., 2016, p. 5).

The final view to mention is the "successful one" in mathematics. This is the so-called formal–abstract view of Kolmogoroff (1933). He explicitly axiomatizes probability theory like Hilbert (1899) does for geometry. This means he explicitly omits an interpretation of the concept of probability and leaves it open. His reason for doing this and not following, for example, von Mises's (1930) approach of inferring probability from a theory of the collective lies in his goal to achieve a "simple formulation of probability" (Kolmogoroff, 1933, p. 2) based on measure theory—a branch of mathematical analysis.

Batanero et al. (2016) summarize that the different views on probability theory provide different opportunities for problem solving and include specific problems in their application. These are the main reasons why probability is a highly discussed concept in mathematics education, mathematics, and mathematical philosophy. This makes probability a paradigmatic mathematical concept in the sense of a "cloud chamber" for thinking about the nature of mathematical concepts and their connection to learning processes, beliefs about mathematics, and also learning biographies in general (see Stoffels, 2020, p. 381).

20.2.3 Why Probability Theory (and Statistics) are Important—A Mathematics Education Research Perspective

Probability theory and statistics are the branches of mathematics where applicability is immediately recognizable. Especially in the twenty-first century, with perceived "data mining," "big data," and "Bayesian networks," probability theory is becoming an increasingly important field for the economy, life science, and even leisure time (cf. Diggle, 2015). Many practical fields—most prominently at the moment, the medical field (see 1.2.2)—make significant use of concepts from probability theory (Batanero et al., 2016).

In addition to its usefulness, the concepts of probability theory give the opportunity for rich concept-building, argumentation, and the integration of disciplines, as it opens discourses between mathematics, philosophy, and physics.

As a last remark from the mathematics education perspective, probability theory still raises questions—as shown in the various examples—about its basic concepts, which influence the learning and teaching of probability and statistics –e.g., what probability is, how to evaluate independence/randomness, etc.

20.3 Physics Education Research

It is impossible to imagine modern physics without probability theory—especially in quantum mechanics, uncertain events are essential components of the theory. In modern descriptions, the position of an electron in the orbit of an atom cannot be determined exactly; only probabilities can be given. However, the usage of probability theory in physics did not appear in the last century with quantum mechanics; thermodynamics also describes the probabilities of events, as in the Maxwell-Boltzmann distribution, which describes the probability density function of the speed of particles at a given temperature (see next section). Finding the influence of probability can be quite difficult—in quantum mechanics and thermodynamics, a great number of particles are often considered. Therefore, based on the law of large numbers, the events can (often, but not always) be described accurately and precisely with simple laws, without perceiving or measuring any random event.

Even in the seemingly precise domain of classical mechanics, uncertainties have to be considered—it is impossible to do any measurement in an experiment, whether at school or in a high-tech lab, without some degree of uncertainty. This uncertainty is important for the connection between experiments and theory: Often, experiments do not provide exactly the results that theory predicts. The exact determination of uncertainties is an important criterion to assess whether the theory (or the experiment) is fundamentally flawed or whether the deviation is due to measurement limitations. In such a case, the experiment may confirm the theory despite a deviation. To determine uncertainties as accurately as possible, statistics are necessary.

At the same time, very little time is spent on the topics of probability theory and statistics in physics classes (e.g., Next Generation Standards 2017 or Kultusministerkonferenz 2020). The corresponding background knowledge largely comes from mathematics lessons, which often leads to a lack of connection between physics and these topics. The extent to which probability theory and statistics are covered in mathematics classes varies significantly depending on the curriculum. Some curricula introduce these topics at an early stage (e.g., Germany, Kultusministerkonferenz 2004), others only in the second half of a student's school career (Langrall, 2018, pp. 39–50, see for example the Common Core State Standards Initiative of 2010, which has been adopted by most states in the United States). The early introduction of probability theory and statistics leads to the better performance of German pupils in the area of probabilities and statistics on the PISA (Sälzer et al., 2013, p. 93). Whether it has an influence on the understanding of probabilities and statistics in physics classes is still to be explored. Even at university, students often have significant problems with the basic concepts of statistics, while their problems with the basic concepts of probability seem to lessen compared to their school days (Mountcastle et al., 2007). So, from a physics education standpoint, there are two main problems: the lack of a mathematical background in probability theory and statistics and their limited (often missing) connections to physics lessons in schools, which makes the transfer from mathematics to physics difficult (Kanderakis, 2016).

To summarize, there are the field of probability theory, which we mostly find in nuclear physics and quantum mechanics, and the field of statistics, which is especially important in the analysis of experiments. Due to the strong connection between the two fields, most of the following points are relevant for both branches.

20.3.1 A Physical Application of Probability Theory: The Speed of Particles

One example that shows the difficulties of seeing random events is the speed of particles in an ideal gas. The speed of atoms and molecules and their temperature are connected. The temperature T leads to the thermal energy E_{Th} of a particle,

$$E_{Th} = \frac{3}{2} k_B T,$$

using the Boltzmann constant k_B. This term is valid for particles that can move in three dimensions (the so-called degrees of freedom).

This can be connected with the classic equation for kinetic energy:

$$E_{kin} = \frac{1}{2}mv^2.$$

This connection provides the well-known relation:

$$v = \sqrt{\frac{3k_BT}{m}}.$$

With this equation, many problems can be solved—for example in astrophysics. But the implicit assertion of this equation—each particle in an ideal gas with a known temperature T has the speed v - is wrong. In reality, the given thermal energy is not the energy that *each* particle has—it is the average energy, a result of the probability distribution of the speed of the particles: the Maxwell-Boltzmann distribution (Fig. 20.2).

As we can see, there is not one speed but a distribution of speeds. The calculated speed v is the root-mean-square speed of the distribution. Even if the distribution of speeds is mentioned in textbooks, it is often not explained that the average value is used (Bader, 2010, pp. 336 ff., Bredthauer, 2019, pp. 320 ff.).

There are several examples like this in physics at school—relations are given without showing their connection to the questions of probability. Because of this lack of connection, many students are not put in a position to see the importance of probabilities in physics.

20.3.2 A Physical Example of Statistics: The Size of a Piece of Paper

The second major area to apply probability theory and statistics is the evaluation of experiments. As mentioned previously, every measurement has a degree of uncertainty, and this has a significant influence on the results of an experiment.

A simple example can clarify the most important steps. To determine the area of a sheet of paper (ISO A4), the two sides must be measured and multiplied together. By measuring the sides l and w several times, we can improve the result using methods of statistics. Each measurement differs a bit from the others, due to small random effects—for example, the measuring tape may slip minimally or the viewing angle when reading may vary slightly. The first step is to

Maxwell-Boltzmann distribution

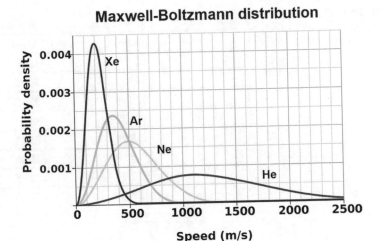

Fig. 20.2 The Maxwell-Boltzmann distribution for Helium, Neon, Argon, and Xenon at the temperature of 298.15 K (public domain)

determine the most probable value: the arithmetic mean.[1] Table 20.3 shows the results of ten measurements of each side, with the arithmetic means $l = 29.7$ cm and $w = 21.0$ cm.

Obviously, the individual values differ from each other. To measure the dispersion of the values, we can determine the standard deviation of the mean. The standard deviations of the mean values are $\sigma_l = 0.02$ cm and $\sigma_w = 0.05$ cm.

Therefore, we can determine the length of the sides a and b:

$$l = 29.7 \text{ cm} \pm 0.02 \text{ cm} \quad w = 21.0 \text{ cm} \pm 0.05 \text{ cm}$$

Thus, we know that the true value of l is between 29.68 and 29.72 cm with a probability of approximately 68%. With a probability of 95%, it lies between 29.66 and 29.74 cm; with a probability of 99%, it lies between 29.64 and 29.76 cm. (For w the deviations are analogous).

The best (most probable) value for the size A of the sheet is the product of the arithmetic means of l and w:

$$\overline{A} = \overline{l} \cdot \overline{w} = 623.7 \text{ cm}^2$$

---

[1] The formulas for calculating the various statistical quantities are not important to understand here and can be found in any textbook on statistics—e.g., Lawrence (2019).

Table 20.3 Measurements
of the sides l and w of an
ISO A4 sheet of paper

Measurement	Side l (cm)	Side w (cm)
1	29.64	21.19
2	29.71	20.98
3	29.75	20.87
4	29.64	21.12
5	29.73	21.02
6	29.79	21.11
7	29.59	20.68
8	29.64	20.85
9	29.79	21.01
10	29.72	21.17

However, due to the uncertainty of the measurements of the sides, the calculated area is also somewhat uncertain. To calculate this, we use the propagation of uncertainty, which is a result of statistical theories:

$$\sigma_A = \sqrt{\left(\frac{\partial A}{\partial l}\right)^2 \cdot \sigma_l^2 + \left(\frac{\partial A}{\partial w}\right)^2 \cdot \sigma_w^2} \approx 1.5 \text{ cm}^2$$

Therefore, the size of the sheet of ISO A4 paper is[2]:

$$A = 623.7 \text{ cm}^2 \pm 1.5 \text{ cm}^2.$$

With a greater number of measurements, the standard deviations of the means of both sides would decrease; thus, the uncertainty of the area would also decrease.

The methods used in this simple example are also the basis of evaluating complex physical experiments and have even found their way into empirical social and educational research, as well as psychology. Still, the mathematical tools used in this section are rarely mentioned in curricula in schools. This is also due to the mathematical requirements such as calculating derivatives (see for example Kultusministerkonferenz 2012 for Germany), which leads to comprehension problems, even in university (Mountcastle et al., 2007).

---

[2] Usually, the arithmetic mean is rounded to the first significant digit of the standard deviation. If the first digit of the standard deviation is one, the next digit is often included because the decimal place makes a large relative difference in the deviation.

20.3.3 Some Typical Individual Concepts

Students already have concepts of physical topics before they start to learn about physics in school. As Vu et al. (2020) show, those concepts (often called misconceptions) should be called individual concepts. Even if they do not correspond to physical theories, they have proven themselves in everyday life for the students. One of the biggest difficulties of using probabilities in physics is that it is often impossible to recognize them. Therefore, for many students (even in university), physics is a totally accurate science; uncertainties that arise during the process of measurement are influenced by humans. This leads to the first individual conception (Hopf & Schecker, 2018, p. 229):

- *In physics, there are no random events.* For many students, any random event is a consequence of our limited knowledge about the initial conditions. If we could know every detail—which is impossible in reality—we would know the exact results, which is exactly what Laplace assumes (see 1.2.2). Actually, this difficulty is not surprising, if we take a look into the history of physics: even Albert Einstein did not like the idea of "true random events," leading to his famous words in a letter to Max Born: "The theory provides much, but it hardly brings us any closer to the mystery of the Old [Author's note: God]. Anyway, I'm convinced he won't roll the dice" (translated from Einstein et al., 1972, p. 97).

This thinking also leads to the next conception (Hopf & Schecker, 2018, p. 230):

- *Identical causes have identical effects.* This is strongly connected to the first misconception. If we have the same situation, the result will be the same. While this (mostly) works in classical mechanics and electrodynamics, in quantum mechanics this principle is not valid. Again, however, in reality, it is a measurement problem: we cannot examine a single photon.

The consideration of single elements causes further difficulties (Hopf & Schecker, 2018, p. 230):

- *The probability of events changes with time.* The typical example of this concept is gambling: while the probabilities in roulette (or any other gambling game) stay the same, everyone has heard the sentence, "There have been three reds in a row; the next one has to be black!" The same idea can be transferred to physics: a radioactive atom that has already had six half-lives should disintegrate rather than one that only had two half-lives. In reality, atoms do not have a memory—they do not "remember" how old they are. Therefore, the probability of decay is the same for each atom.

These individual conceptions show some problems students encounter when dealing with probability theory in physics. Especially, the idea that physics is a science that deals only with determined processes leads to problems of understanding—the aspect that some phenomena in physics cannot be measured or calculated exactly in a deterministic way can be hard to accept. These concepts also reveal a second point: we can divide the main difficulties that students have with probabilities and statistics in physics into two main groups. There are problems with the basic understanding and calculations of topics that are quite similar to the problems in mathematics (the third concept) and there are problems about the nature of physics (the first two concepts).

20.3.4 Why Probability Theory and Statistics are Important—A Physics Education Research Perspective

We have seen a significant lack of probability theory and statistics in physics (and mathematics) lessons—but *why* should these topics be addressed? It is possible to calculate many things without thinking about probabilities.

However, this is not true for substantial aspects of twentieth-century physics. We have already seen that it is impossible to understand modern physics without probability theory and statistics. Probability theory is an essential part of quantum mechanics, and nuclear physics and thermodynamics are also based on probabilities. Physics without probabilities would therefore be on par with the physics of the early nineteenth century. Furthermore, it is impossible to conduct any experiments without statistics. The evaluation of any experiment (small experiments in the physics lab as well as projects at the Large Hadron Collider at CERN in Switzerland) requires statistics. These topics are not only relevant for physicists, though.

In addition, many careers are highly connected to statistics, as mentioned in Sect. 1.2.4. Empirical social science is based on statistical methods, including psychology and educational research. Their methods are based on the same fundamentals as the evaluation of physical experiments, even if the focus differs. In any profession that deals with big data, statistical methods must be applied.

The handling of the corresponding methods is therefore relevant in many academic disciplines, accordingly, the basics should be practiced in school if possible.

Regardless of profession, both probability theory and statistics are relevant for another reason: they are an essential part of the modern worldview. In order to understand the nature of science– for example, the methods of epistemology and the limitations of measurement techniques—a basic understanding of both topics is essential. Probability theory and statistics are an important part of scientific literacy. In daily life, as well, we have to deal with probabilities and statistics, often in ways that are closely related to physics. One famous example that shows the misunderstanding of probabilities is radioactivity. Many people are afraid of small amounts of radiation, not knowing that there is background radiation in daily life. While there are profound risks from high doses of ionizing radiation (which people are aware of), there are problems with the assessment of low doses of radiation, which leads to unnecessary fears (Wojcik et al., 2018). In order to take part in a rational discussion about this (socially quite relevant) topic, however, one must be guided not by fears but by facts—and many of these facts are based on statistics.

Also, due to the COVID-19 pandemic in 2020/2021, we hear many statistical terms. Often, the statistics are hidden or greatly simplified in the news (for example, the basic reproduction number of COVID-19). To understand these terms and their implications, knowledge of statistics is necessary. Actually, the methods used in physics also can be applied to epidemics, which offers the opportunity to work with students on a highly up-to-date topic (MacIsaac, 2020).

Those are just two examples that show how important it is to understand the basics of probability theory and statistics—not only for pure physics but also for daily life. Ultimately, the reasons why probability theory and statistics are important from the perspective of physics education are quite close to the perspective of mathematics education (despite the parts that are closely related to pure physics).

20.4 Summary

To summarize, there has been much research in mathematics education on the concepts of probability and statistics, while the topics are less explored in physics education, especially secondary school physics. In both disciplines, the special nature of probability and its connection to data measurement, analysis, and interpretation are the roots of students' difficulties.

In politics, there was a change regarding mathematics education in the early 2000s, moving towards broader implementation of statistics and probability theory in school curricula. However, there is still a great disparity between the significance of probability theory and statistics for physics and daily life and the time spent in class on these topics.

Even nowadays, it seems that probability theory and statistics are not the focus of mathematics teaching and learning at school, so that Batanero et al. conclude:

"However, including a topic in the curriculum does not automatically assure its correct teaching and learning; the specific characteristics of probability, such as a multifaceted view of probability or the lack of reversibility of random experiments, are not usually found in other areas and will create special challenges for teachers and students."
(Batanero et al., 2016, p. 1)

Due to these mathematical challenges, students do not have the necessary background knowledge to apply the concepts in physics, and it is difficult to "catch them up" in physics class. Therefore, the aspects of physics containing probabilities and statistics are often avoided or simplified. This leads to a gap between secondary school and university, where statistics, in particular, is an essential part of many courses—not just in natural science but also the social sciences.

However, even in university, many students cannot overcome their deficits in the necessary topics and need mathematical support to understand the foundational theories, conceptualizations, and their application to physics. If these issues are never addressed and these students become teachers, then they are likely to have crucial problems teaching the topics—or they will avoid them if possible, which will lead to a vicious circle.

References

Bader, F. (2010). *Physik. Gymnasium SEK II [Physics. Gymnasium Secondary Level II].* Schroedel.

Batanero, C., Chernoff, E. J., Engel, J., Lee, H. S., & Sánchez, E. (2016). Research on teaching and learning probability. Springer International Publishing (ICME-13 Topical Surveys).

Backhaus, I., Dragano, N., Boege, F., Göbels, K., Hermsen, D., Lübke, N., & Timm, J. Seroprävalenz COVID-19 Düsseldorf: SERODUS I & II. Feldbericht und vorläufiger Ergebnisbericht [Seroprevalence COVID-19 Düsseldorf: SERODUS I & II. Field report and preliminary results report]. Universitätsklinikum Düsseldorf; Heinrich Heine Universität Düsseldorf. https://www.uniklinik-duesseldorf.

de/fileadmin/Fuer-Patienten-und-Besucher/Kliniken-Zentren-Institute/Institute/Institut_fuer_Medizinische_Soziologie/Forschung/SeroDus/Feld-_und_Ergebnisbericht_SERODUS-I_SERODUS-II_03-02-2021_v01.pdf.

Ben-Zvi, D., & Garfield, J. (2004). *The challenge of developing statistical literacy, reasoning and thinking.* Springer.

Bernoulli, J. (1899). Wahrscheinlichkeitsrechnung (Ars conjectandi), übersetzt von R. Haussner [Probability Theory (Ars conjectandi). Translated by R. Haussner]. Ostwalds Klassiker der exakten Wissenschaften, 107.

Biehler, R., & Engel, J. (2015). Stochastik: Leitidee Daten und Zufall [Stochastics: central idea data and chance]. Springer. http://link.springer.com/chapter/https://doi.org/10.1007/978-3-642-35119-8_8/fulltext.html.

Bredthauer, W. (2019). *Impulse Physik. Oberstufe [Impulse Physics Secondary Level II].* Klett.

Carnap, R. (1950). *Logical foundations of probability.* University of Chicago Press.

Chernoff, E. J., & Sriraman, B. (2014). *Probabilistic thinking.* Springer, Netherlands.

Common Core State Standards Initiative. (2010). Common Core State Standards for Mathematics. http://www.corestandards.org/wp-content/uploads/Math_Standards1.pdf.

Czuber, E. (1898). Die Entwicklung der Wahrscheinlichkeitstheorie und ihrer Anwendungen [The development of probability theory and its applications]. Leipzig (Jahresbericht der Deutschen Mathematiker-Vereinigung).

Deutsches, P. K., Baumert, J., Artelt, C., Klieme, E., Neubrand, M., Prenzel, M., ... & Weiß, M. (Eds.) (2013). PISA 2000—Die Länder der Bundesrepublik Deutschland im Vergleich [PISA 2000-The States of the Federal Republic of Germany in Comparison]. Springer.

Diggle, P. J. (2015). Statistics: A data science for the 21st century. *Journal of the Royal Statistical Society: Series A (statistics in Society), 178*(4), 793–813. https://doi.org/10.1111/rssa.12132.

Einstein, A., Born, H., & Born, M. (1972). *Briefwechsel 1916–1955 [Correspondence 1916–1955].* Rowohlt Taschenbuchverlag.

Freudenthal, H. (1963). Wahrscheinlichkeit und Statistik [Probability and Statistics]. R. Oldenbourg.

Hilbert, D. (1899). *Grundlagen der Geometrie [Basics of Geometry].* B. G. Teubner.

Hilbert, D. (1902). Mathematical problems. In Bull. *Journal of the American Mathematical Society, 8*(10), 437–480. https://doi.org/10.1090/S0002-9904-1902-00923-3.

Hopf, M., & Schecker, H. (2018). Schülervorstellungen zu fortgeschrittenen Themen der Schulphysik [Student presentations on advanced topics in school physics.]. In H. Schecker et. al. (eds.), Schülervorstellungen und Physikunterricht. Ein Lehrbuch für Studium, Referendariat und Unterrichtspraxis [Student conceptions and physics education. A textbook for studies, teacher training and teaching practice.] (pp. 225–242). Springer Spektrum.

Keynes, J. M. (1921). *A treatise on probability.* MacMillan.

Kanderakis, N. (2016). The mathematics of high school physics. *Science & Education, 25,* 837–868. https://doi.org/10.1007/s11191-016-9851-5.

Kultusministerkonferenz. (15.10.2004). Bildungsstandards im Fach Mathematik für den Primarbereich (Jahrgangsstufe 4). KMK-Bildungsstandards Mathe Primar [Educational standards in mathematics for the primary level (grade 4). KMK Educational Standards in Primary Mathematics], S. 1–35. https://www.kmk.org/fileadmin/Dateien/veroeffentli-chungen_beschluesse/2004/2004_10_15-Bildungsstandards-Mathe-Primar.pdf.

Kultusministerkonferenz. (18.06.2020). Bildungsstandards im Fach Physik für die Allge-meine Hochschulreife [Educational standards in physics for the general university entrance qualification]. https://www.kmk.org/fileadmin/Dateien/veroeffentlichungen_beschluesse/2020/2020_06_18-BildungsstandardsAHR_Physik.pdf.

Kolmogoroff, A. (1933). Grundbegriffe der Wahrscheinlichkeitsrechnung, Ergebnisse der Math, u. Grenzgebiete [Basic concepts of probability, results of math, and related fields.]. Springer.

Kolmogorov, A. N. (1956). *Foundations of the theory of probability: Translation edited by N.* Chelsea Publishing Company.

Langrall, C.W. (2018). The status of probability in the elementary and lower secondary school mathematics curriculum: The rise and fall of probability in school mathematics in the United States. In C. Batanero & E. Chernoff (eds.), *Teaching and Learning Sto-chastics.* ICME-13 Monographs (pp. 39–50). Springer.

Laplace, P. S. (1986). Essai philosophique sur les probabilités [Philosophical essay on Probabilities]. Christian Bourgois (Original work, published in 1814).

Lampinen, J., & Vehtari, A. (2001). Bayesian approach for neural networks—Review and case studies. *Neural Networks, 14*(3), 257–274. https://doi.org/10.1016/S0893-6080(00)00098-8.

Lawrence, A. (2019). *Probability in physics. Undergraduate lecture notes in physics.* Springer.

MacIsaac, D. (2020). Adapting simple mechanics numeric computer modeling to epidem-ics like coronavirus. *The Physics Teacher, 58,* 286. https://doi.org/10.1119/1.5145487.

Mountcastle, D., Bucy, B., & Thompson, J. (2007). Student estimates of probability and uncertainty in advanced laboratory and statistical physics courses. *AIP Conference Pro-ceedings, 951,* 152. https://doi.org/10.1063/1.2820919.

National Reasearch Council. (2017). Next Generation Science Standards. https://www.nextgenscience.org/sites/default/files/AllDCI.pdf.

NCTM. (2000). Principles and standards for school mathematics. VA.

Popper, K. R. (1959). The propensity interpretation of probability. *The British Journal for the Philosophy of Science, 10*(37), 25–42.

Ramdas, K., Darzi, A., & Jain, S. (2020). 'Test, re-test, re-test': Using inaccurate tests to greatly increase the accuracy of COVID-19 testing. *Nature Medicine, 26,* 810–811. https://doi.org/10.1038/s41591-020-0891-7.

Sälzer, C., Reiss, K., Prenzel, M., Schiepe-Tiska, A., & Heinze, A. (2013). Zwischen Grundlagenwissen und Anwendungsbezug: Mathematische Kompetenz im interna-tionalen Vergleich [Basic knowledge and real-life context: International comparison of mathematics competencies]. In M. Prenzel, C. Sälzer, E. Klieme, & O. Köller (Eds.), *PISA 2012: Fortschritte und Herausforderungen in Deutschland [PISA 2012: Progress and challenges in Germany]* (pp. 47–97). Waxmann.

Seifried. J., Böttcher, S., Oh, D. Y., Michel, J., Nitsche, A., Jenny, M. A., Wieler, L. H., Antão, E.-M., Jung-Sendzik, T., Dürrwald, R., Diercke, M., Haas, W., Abu Sin, M., Eckmanns, T., Hamouda, O., & Mielke, M. (2021). Was ist bei Antigentests zur Eigenanwendung (Selbsttests) zum Nachweis von SARS-CoV-2 zu beachten? [What should be considered when using antigen tests for self-testing to detect SARS-CoV-2?] *Epid Bull, 8,* 3–9 https://doi.org/10.25646/8040.

Stoffels, G. (2020). (Re-)Konstruktion von Erfahrungsbereichen bei Übergängen von empirisch-gegenständlichen zu formal-abstrakten Auffassungen. Eine theoretische Grundlegung sowie Fallstudien zur historischen Entwicklung der Wahrscheinlichkeitsrechnung und individueller Entwicklungen mathematischer Auffassungen von Lehramtsstudierenden beim Übergang Schule-Hochschule [(Re-)construction of fields of experience during transitions from empirical-objective to formal-abstract understandings. A theoretical foundation as well as case studies on the historical development of probability theory and individual developments of mathematical conceptions of student teachers during the transition from school to university]. Universi.

Trevena, L. J., et al. (2013). Presenting quantitative information about decision outcomes: A risk communication primer for patient decision aid developers. *BMC Medical Informatics and Decision Making, 13,* S7. https://doi.org/10.1186/1472-6947-13-S2-S7.

Wojcik, A., Hamza, K., Lundgård, I., Enghag, E., Haglund, K., Arvanitis, L., et al. (2018). Educating about radiation risks in high schools: Towards improved public understanding of the complexity of low-dose radiation health effects. *Radiation and Environmental Biophysics, 58,* 13–20. https://doi.org/10.1007/s00411-018-0763-4.

Lesson Plan: Probabilities and Statistics 21

Nguyen Phuong Chi

Lesson Title: Using statistics to determine the free fall acceleration of an object

Abstract: The central element of these two lessons is conducting an experiment to deter-mine the time it takes an object to free fall from various heights. Through this activity, students have the opportunity to review essential mathematics and physics knowledge. To do so, students perform an experiment to determine the time it takes an object to free fall using the devices typically used in physics. In doing so, they apply their knowledge of statistics from mathematics.

Type of school / Grade	High School / 10
Prerequisites	Mathematics – statistical data sequences; mean, median, and mode Physics – concept and properties of a free fall – formula for the free fall acceleration of an object – calculation of error measurement
Number of periods	2

N. P. Chi (✉)
Hanoi National University of Education,
Faculty of Mathematics and Informatics, Hanoi, Vietnam
e-mail: chinp@hnue.edu.vn

Objectives	Mathematics: – Students can establish statistical data sequences – Students can calculate the mean of a statistical data sequence and understand its meaning. Physics: – Students understand the definition and properties of a free fall and the formula for free fall acceleration – Students perform an experiment measuring the time it takes an object to free fall from different heights using an electronic timer – Based on the experimental data, students can determine the average free fall acceleration of an object and calculate the measurement error

21.1   Methodical Commentary

Before conducting the experiment (Phase 3), it is important to clearly demonstrate how to use the measuring device: the electronic timer, its related components, and its principles of operation. Model slowly and clearly how to set up and perform the experiment. At the end of this activity, each group of students must have a table with the times it takes an object to free fall from different distances.

In the activity in Phase 4, students should be encouraged to combine mathematics and physics knowledge to solve the problem. More specifically, they should use the formula for mean from statistics to calculate the average falling time $\overline{t_i}$ and the average squared falling times $\overline{t_i^2}$ at each distance $s_i (i = 1, 2, 3, 4)$. Then they should apply the formula for free fall acceleration: $g_i = \frac{2s_i}{\overline{t_i^2}} (i = 1, 2, 3, 4)$. After that, they need to apply the mean formula again to find the average free fall acceleration:

$$\overline{g} = \frac{g_1 + g_2 + g_3 + g_4}{4}$$

Lastly, they have to calculate the measurement error and write the result:

$$\Delta g_1 = |\overline{g} - g_1|, \Delta g_2 = |\overline{g} - g_2|, \ldots g = \overline{g} \pm (\Delta g)_{max}.$$

21.2 Lesson Plan

No./Time	Stage	Learning activities	Interaction form	Materials/Resources	Methodological comments
# 1 5 min	Beginning the lesson	Say hello to students and tell them the main content of the lesson: they will perform an experiment measuring the time it takes an object to free fall from different heights using an electronic timer and determine the free fall acceleration of the object	Teacher presents	Board, projector	
# 2 20 min	Motivation	Explain to students that this lesson requires the integration of mathematics and physics knowledge they have previously learned. Then ask them to complete **Worksheet 1** to review the necessary concepts	Teacher presents/individual work	Worksheets	Prior knowledge to be reviewed: – Mathematics: statistical data sequences; mean, median, and mode – Physics: the concept and properties of a free fall, the formula for the free fall acceleration of an object, and the calculation of measurement error
# 3 30 min	Practice	Divide students into groups. Tell them that they are performing an experiment to measure the time it takes an object to free fall from different distances using electronic timers. Model for students how to set up and perform the experiment. Explain that they need to measure the time it takes the object to free fall from the heights 0.050m, 0.200m, 0.450m, and 0.800m, and write the results in Table 21.1 on **Worksheet 2**. Students perform the experiment in groups. After they finish, invite one or two groups to present and explain their results	Teacher presents/Group work	Electronic timers and related components, falling objects, worksheets	– Make sure to explain the measurement clearly – At the end of this activity, each group must have a table of the free fall times of the object from different heights

No./Time	Stage	Learning activities	Interaction form	Materials/Resources	Methodological comments
# 4 30 min	Application	Ask students to present how to determine the free fall acceleration of the object based on the statistical data table (Table 21.1) from their experiment. Suggest that they use the content they reviewed with **Worksheet 1**. Ask them to add three more columns: $\bar{t_i}$ (the average falling time at the height s_i), $\bar{t_i^2}$ (the average squared falling time at the height s_i), and g_j (the free fall acceleration at the height s_i), where $i = 1, 2, 3, 4$ After that, students calculate the corresponding values in each column (fill in Table 21.2 on **Worksheet 2**). Ask students to calculate the average free fall acceleration of the object and the measurement error. write the results. After they finish, invite one or two groups to present and explain their results	Teacher presents/group work	Worksheets	– A combination of math and physics knowledge is required to solve the problem – Calculate $\bar{t_i}$ and $\bar{t_i^2}$ at for each height – Apply $g_i = \frac{2s_i}{\bar{t_i^2}}$ for the free fall acceleration – Apply the formula for $\bar{g}$ to find the average free fall acceleration – Calculate the measurement error and write the result
# 5 5 min	Ending the lesson	Summarize what students have done and learned in the lesson. Emphasize that statistics can be applied in real life and many other fields, such as physics. Say goodbye to students	Teacher presents	Board, projector	

Worksheet 1

Name of student: ....

Class: ....

Please answer the following questions:

1. What is a free fall? Describe the properties of a free fall. Write the formula for the free fall acceleration of an object.

   ....

   ....

   ....

   ....

   ....

   ....

   ....

2. Define the following terms related to statistical data sequences: mean, median, mode. When we measure a quantity A n times, we receive the statistical data sequence $A_1, A_2, \ldots, A_n$. Which measure (mean, median, or mode) should be chosen to represent this sequence? Explain your answer.

   ....

   ....

   ....

   ....

   ....

   ....

   ....

3. Describe how to determine the measurement error for Question 2.

....

....

....

....

....

....

Worksheet 2

Group number: ....

Class: ....

1. Perform the experiment, measuring the time it takes an object to free fall from the following heights: 0.050 m, 0.200 m, 0.450 m, and 0.800 m. Then, write the measured times in the table below:

2. Describe how to determine the free fall acceleration of the object based on the statistical data table (Table 21.1). What mathematics and physics knowledge should be applied?

....

....

....

....

....

....

....

3. Adding three more columns to the table: $\bar{t_i}$, $\bar{t_i^2}$, g_i to the table 21.1, we obtain Table 21.2 below. Please calculate the corresponding values in each column and fill in Table 21.2.

Table 21.1 Free fall times of an object from different heights Starting position of the falling object: $s_0 = \ldots$ (m)

Distance s(m)	Falling time t (s)				
	1st fall	2nd fall	3rd fall	4th fall	5th fall
0.050					
0.200					
0.450					
0.800					

Table 21.2 Investigating a free fall Starting position of the falling object: $s_0 = \ldots$ (m)

Distances(m)	Falling time t(s)					$\overline{t_i}$	$\overline{t_i^2}$	$g_i = \frac{2s_i}{t_i^2}$
	1st fall	2nd fall	3rd fall	4th fall	5th fall			
0.050								
0.200								
0.450								
0.800								

4. From the data in Table 21.2, please calculate the average free fall acceleration of the object and the measurement error and write the results.

Comparison: Light Rays and the Intercept Theorem in Mathematics and Physics Education

Frederik Dilling and Ina Stricker

22.1 Introduction

This chapter introduces the intercept theorem in mathematics education and its application for the description of light rays in physics education. This application is one of the classical content-related connections between mathematics and physics. Also, the first approaches of the intercept theorem were likely developed from a practical problem—the determination of the height of a pyramid. Thus, according to the writings of Diogenes Laertius, Thales of Miletus (624/623–548/545 BC) supposedly gained the following insight:

> Hieronymus reports that he measured the height of the pyramids by means of their shadow, which he measured at the exact moment when our shadow and our body have the same length. (Gericke, 1984, S. 75, English translation)

Even if the described method does not seem suitable for measuring the height of a pyramid (too flat angle of inclination, specific position of the sun necessary)

F. Dilling (✉)
Mathematics Education, University of Siegen, Siegen, Germany
e-mail: dilling@mathematik.uni-siegen.de

I. Stricker
Physics Education, University of Siegen, Siegen, Germany
e-mail: stricker@physik.uni-siegen.de

© The Author(s), under exclusive license to Springer Fachmedien Wiesbaden GmbH, part of Springer Nature 2022
F. Dilling and S. F. Kraus (eds.), *Comparison of Mathematics and Physics Education II*, MINTUS – Beiträge zur mathematisch-naturwissenschaftlichen Bildung, https://doi.org/10.1007/978-3-658-36415-1_22

(cf. Gericke, 1984), it shows a first conception of similarity and the connection between geometric optics and the intercept theorem. In the following sections, this topic will be examined from the perspectives of mathematics and physics.

22.2 Mathematics Education Research

22.2.1 Similarity and the Intercept Theorem

The concept of similarity is used in geometry to investigate the properties of figures that have equal angles and equal ratios of the lengths of corresponding sides (see Fig. 22.1). Two approaches to the concept of similarity can be distinguished in geometry. On the one hand, similar polygons can be defined by equal inner angles and corresponding ratios of sides. On the other hand, figures can be defined as similar if there is a similarity mapping between them, which is composed of a centric dilation and a congruence mapping (e.g., rotation and reflection). Since the first approach is related to the intercept theorem, similarity mappings will not be discussed further in this chapter.

In school mathematics education, the concept of similarity is often used in the context of triangles (cf. Helmerich & Lengnink, 2016). It is possible to formulate so-called similarity theorems that identify similar triangles based on certain properties (see Fig. 22.2). Accordingly, two triangles are similar to each other if they:

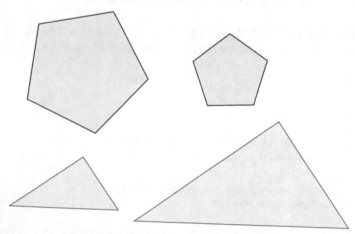

Fig. 22.1 Similar triangles and pentagons. (Drawn with ©GeoGebra)

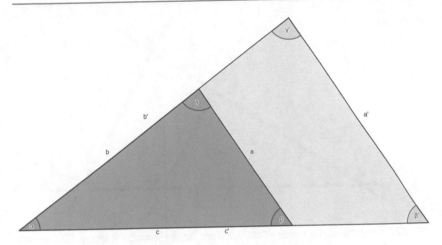

Fig. 22.2   Similar triangles in front of each other. (Drawn with ©GeoGebra)

- match in all their angles *or*
- match in all ratios of corresponding sides *or*
- match in an angle and the ratio of the adjacent sides *or*
- match in the ratio of two sides and the opposite angle of the larger side.

The equality of the ratios of the sides in similar triangles leads to the intercept theorem. This theorem allows the identification of the similarity of two triangles in specific positions (see Fig. 22.3 and 22.4) or the determination of undetermined side lengths from other known side lengths.

Intercept Theorem: Suppose two lines[1] intersect at point Z and two other parallel lines do not coincide with Z. The intersections of the parallels with the first and the second line should be named A and A' resp. B and B'. Then, the following statements are true:

1. The ratios of any two segments on the first line equal the ratios of the corresponding segments on the second line:

   a) $\frac{|ZA|}{|ZA'|} = \frac{|ZB|}{|ZB'|}$ b) $\frac{|ZA|}{|AA'|} = \frac{|ZB|}{|BB'|}$ c) $\frac{|ZA'|}{|AA'|} = \frac{|ZB'|}{|BB'|}$

---

[1] For a discussion dealing with lines in mathematics education, see Struve (1990).

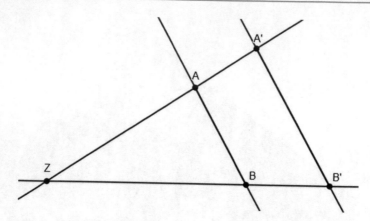

Fig. 22.3 Possible constellation of straight lines for the intercept theorem. (Drawn with ©GeoGebra)

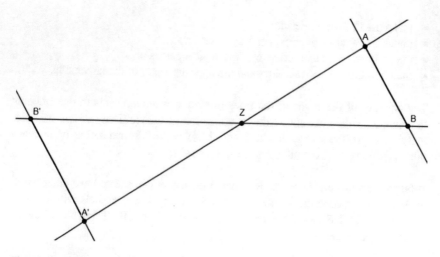

Fig. 22.4 Another possible constellation of straight lines for the intercept theorem. (Drawn with ©GeoGebra)

2. The ratio of the two segments on the same line starting at point Z equals the ratio of the segments on the parallels:

a) $\frac{|ZA|}{|ZA'|} = \frac{|AB|}{|A'B'|}$ b) $\frac{|ZB|}{|ZB'|} = \frac{|AB|}{|A'B'|}$

The converse statement to the intercept theorem is only valid for its first part. For example, if $\frac{|ZA|}{|ZA'|} = \frac{|ZB|}{|ZB'|}$, then the two line segments AB and $A'B'$ are parallel to each other. However, if $\frac{|ZB|}{|ZB'|} = \frac{|AB|}{|A'B'|}$, they do not necessarily have to be parallel to each other (see Fig. 22.5).

22.2.2 Pantograph: A Historical Drawing Instrument

A pantograph is a tool for copying, scaling up and scaling down drawings (see Fig. 22.6) and is a classical application of the intercept theorem in school. The drawing tool is constructed as follows (for the inscriptions, see the schematic drawing in Fig. 22.7). The point Z, also called the pole, is fixed on the base, e.g. with a screw. At point B, a tracing pen is attached and used to trace a given figure on paper. The figure traced with B is mapped onto a figure drawn with a pen at point B' by the mechanism of the tool. The drawing pen sketches an upscaled figure similar to the original figure. The scale factor is given by the ratio $|\overline{ZB'}|:|\overline{ZB}|$. This scale factor can be changed by adjusting the rails at points A and C. The respective stretch factor is marked directly on the rails on most

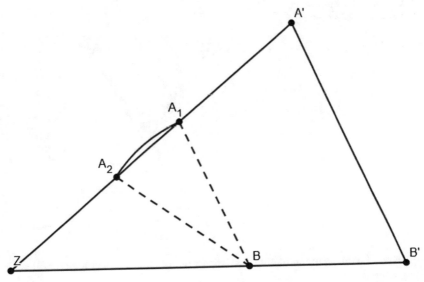

Fig. 22.5 Illustration of the converse statement of the second part of the intercept theorem. (Drawn with ©GeoGebra)

Fig. 22.6 3D-printed pantograph. (Dilling et al., 2020)

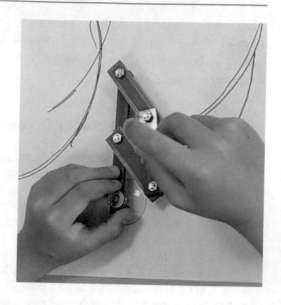

Fig. 22.7 Schematic illustration of a pantograph. (Drawn with ©GeoGebra)

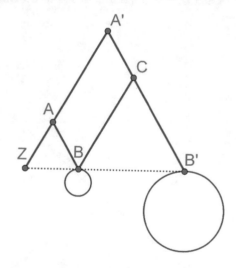

devices. If you swap the tracing pen and drawing pen, the pantograph will scale down the figure.

The instrument is based on the mathematical concept of centric dilation. The construction consists of a flexible linkage with four rails. Figure 22.7 shows a schematic illustration of a pantograph. The four rails correspond to the line segments ZA', $A'B'$, AB and BC. The rails ZA' and $A'B'$ have the same length so that $ZB'A'$ corresponds to an isosceles triangle. In addition, the rails with sections AB and BC are bolted to the other rails so that they form a flexible parallelogram $(ABCA')$ and Z, B and B' are collinear. Thus, the triangles ZBA and $ZB'A'$ are always similar. The mapping is a centric dilation with the center Z. The same applies to the figures drawn with B and B'.

The intercept theorem can be applied to the two triangles of the pantograph: ZBA and $ZB'A'$. According to the first part of the theorem, the ratio $|\overline{ZB'}| : |\overline{ZB}|$ is equal to the $|\overline{ZA'}| : |\overline{ZA}|$. This ratio represents the scale factor of the drawn figures and can be denoted by k. With a ratio of $\frac{|\overline{ZA'}|}{|\overline{ZA}|} = \frac{3}{1}$ between the lengths of the line segments, $k = 3$ and, thus, the point B' shifts by three units of length if the point B is shifted by one unit. Thus, the figure is scaled up by a factor of 3. In mathematics class, the first step is to examine the specific case in which the articulation of the pantograph at points A and C is fixed exactly in the middle of the bars. Thus, $|\overline{ZA'}| = 2|\overline{ZA}|$, which means that the scale factor is $k = 2$. The drawing figure is doubled compared to the original figure. If the tracing pen and the drawing pen are swapped, the new scale factor corresponds to the reciprocal value of the previous factor.

The pantograph, with its geometric and technical characteristics, offers a rich variety of applications in mathematics lessons. Through instrumental activities with the drawing instrument, even elementary school students can derive initial characteristics of the construction and functionality of the instrument. Appropriate learning environments must be designed for the subject matter and the pupils (e.g. Dilling et al., 2020). A decisive insight that students can make with the pantograph is that the points Z, B and B' always lie on a straight line. In addition, they can discover that the enlargement or reduction ratio can be read not only from the drawing but also from the device itself. In middle school and high school, students can then investigate the mathematical background completely (cf. Dilling & Vogler, 2020).

22.2.3 Determination of Non-Accessible Distances

In the context of the intercept theorem, mathematics textbooks offer a variety of practical problems. These usually involve the measurement of inaccessible distances, such as the width of a river or the height of a building. This type of task will now be carried out with methods by which foresters can determine the height of trees (cf. Hölzl, 2018).

To determine the height of a tree standing perpendicular to a flat surface, a stick of known length is placed perpendicular to the surface so that the shadow of the tip of the stick just disappears in the shadow of the tree. The light rays fall on the ground at the same angle, and both the tree and the stick are perpendicular to the ground. Therefore, in a simplified sketch (Fig. 22.8), two similar triangles can be identified. According to the intercept theorem, the height of the tree (t) corresponds to the height of the stick (s) in the same way as the distance between the tree and the tip of the shadow (d_t) corresponds to the distance between the stick and the tip of the shadow (d_s). Therefore, the height of the tree can be calculated from the distances on the ground and the height of the stick: $t = s\frac{d_t}{d_s}$.

Another method uses a simple tool called the forester's triangle (German: Försterdreieck) (cf. Helmerich & Lengnink, 2016) (Fig. 22.9). This is an isosceles triangle with one right angle. The location of an observer is chosen in such a way that he can aim the hypotenuse of the right-angled triangle at the tip of the

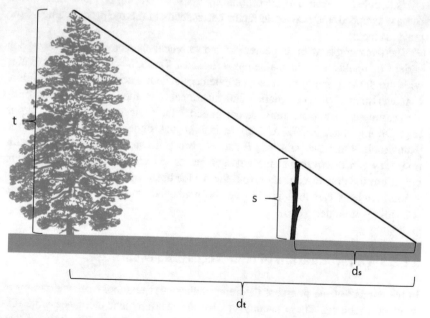

Fig. 22.8 Measuring the height of a tree with a stick. (© Frederik Dilling)

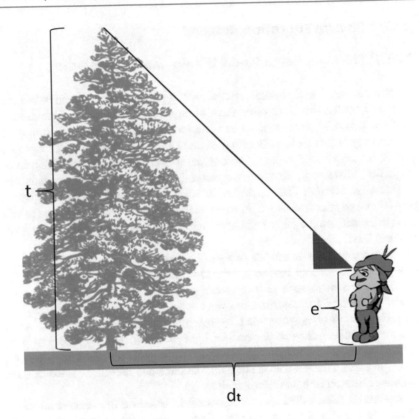

Fig. 22.9 Measuring the height of a tree with a triangle (© Frederik Dilling)

object. It is crucial that the triangle is placed with one cathetus horizontal and the other vertical and perpendicular to the ground. Since it is an isosceles triangle, the horizontal distance between the tree and the triangle (d_t) corresponds exactly to the vertical distance between them (intercept theorem). To calculate the height of the tree (t), the observer only has to add the eye level (e): $t = d_t + e$.

If a right-angled, non-isosceles triangle is used, the distance to the tree must be additionally multiplied by the ratio of the two catheti.

As an alternative to the intercept theorem, it is also possible to use basic trigonometry to calculate the height of a tree. This approach will not be discussed in this chapter.

22.3 Physics Education Research

22.3.1 The Light Beam Model Within Geometrical Optics

In optics, the section of physics that deals with light, its properties and its interactions with matter, similar to other areas of physics, different models are used. A model is generally understood as an ideal (conceptual) or a material (representational) object that replaces the original and can also be viewed as a simplification of the original.[2] Although, the original and the model do not match in all properties. Models are, therefore, only useful and reasonable within certain limits (Meyer & Schmidt, 2003). Kircher (2010) describes the function of models in physics and physics teaching as follows: on the one hand, models are used to gain knowledge and, on the other hand, they help students keep, reproduce and apply physical facts.

There are three basic models of light: light rays, light waves and light quantum. Likewise, different models are used to explain light phenomena. For example, light phenomena such as interference patterns are explained using the wave model of light, which describes the wave nature of the light. Alternatively, the photoelectric effect is interpreted by using the light quantum model, which describes light as a flow of photons (or light particles). The ray model of light, which illustrates its straight-line propagation, is generally used in secondary school to explain the phenomena of light refraction and reflection (a deduction of the wave character of light is also possible).

Ray optics (also called geometric optics) is based on the conceptual model of the "light ray" and the premise that light spreads straight in a homogeneous medium. As previously mentioned, phenomena such as light reflection and refraction at the boundary of two media can be explained by and light paths can be constructed with optical components and devices. Although, this procedure neither describes the nature of light nor does it reveal anything about light's character.

The roots of the concept of the light beam likely lie in antiquity. Pythagoras (570 BC–510 BC) explained the visual process through the so-called visual rays and presumably laid the basis for this idea. The works of Euclid (third century BC) serve as the foundation of geometric optics in its current form (Weinmann, 1980).

---

[2] See also Volume 1, Chapter "Models and Modeling" (Tran, Chu, Holten, Bernshausen, 2020).

As a rule, the imagination of light rays does not pose any difficulties for learners. Although this model is a theoretical construct, its introduction should be related to phenomena. The term "light beam" only becomes meaningful if it is developed in connection with specific natural phenomena such as the creation of shadows (Jung, 1979). Following the introduction of the model and the first constructions of light beams, the teacher should address an important aspect of physics education, namely the distinction between phenomena that can be experienced and the models that are used to explain the phenomena. An example will illustrate this educational objective: light rays of the sun, which become visible between the trees in the forest. For example, light rays in fog (Tyndall effect) are part of an experienceable phenomenon that can be explained by light scattering, whereas light rays that are used to explain the formation of a specular reflection have a model character. Thus, the model character of the notion of light rays in geometrical optics shall be clearly emphasized (Haagen-Schützenhöfer & Hopf, 2018).

The light beam model as well as the geometrical constructions of geometrical optics may appear relatively uncomplicated to teachers. However, research in physics education indicates that a premature introduction to light beam constructions with only two or three distinct light beams (parallel ray, central ray, focal ray, Fig. 22.10) leads to misconceptions from the pupils. For example, in the

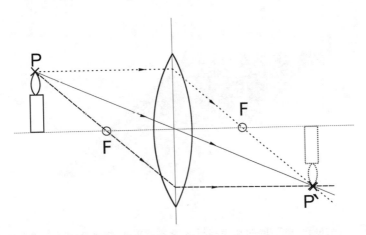

Fig. 22.10 Formation of an image on a converging lens, construction based on three selected light rays: parallel ray—dotted line, central ray—solid line, focal ray—dashed line; F: focal point of the converging lens; P: a point of the object and P': a point of the image (according to Meyer, 2003, p. 329)

learner's imagination, it is not possible to completely visualize an object with the help of a lens if the object is larger than the diameter of the lens. For this reason, Haagen-

Schützenhöfer & Hopf (2018) recommend first working with the idea of a light cone (Fig. 22.11) and incorporating the light beam constructions with two or three distinct light beams into the class later.

Concerning the graphical representation of a light beam in physics, there are clear differences in how the geometrical term "beam" or "semi-straight line" is defined. The starting point of a light beam is not marked by a point like that of a beam in geometry; the light beam in physics starts at the light source. Arrows are used to show the direction of the light rays. Both light beams in physics and beams or semi-straight lines in geometry end in infinity. This characteristic is not depicted in the graphic representation; it must be discussed separately with learners in physics and math lessons to avoid misconceptions. Haagen-Schützenhöfer and Hopf (2018) point out that graphic representations from textbooks are constantly reviewed for potential sources of misconceptions by teachers. In some textbooks, for example, there are overly reduced representations in which the light rays end in their focus after passing through the converging lens.

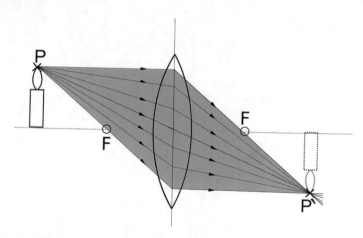

Fig. 22.11 Construction based on a light cone with many light rays: A light cone starts from a point P of the object and leads to the image point P` due to the light refraction at the converging lens (according to Meyer, 2003, p. 329)

22.3.2 Scattering of Light and the Role of Student Perceptions in Physics Education

The students' pre-instructional ideas form the starting point for teaching and learning processes. In didactic literature, everyday terms or misconceptions are often used synonymously with the term "preschool student perceptions" (Duit, 2004; Vu et al., 2020). Teachers should be aware of what learners think about phenomena because in most cases learners' ideas do not match physical ideas and are very persistent against change. For example, pupils often think that a flat mirror swaps left and right, that the reflected image lies on the mirror surface or that the diameter of a lens determines the size of the image (Haagen-Schützenhöfer & Hopf, 2018).

Lack of understanding of a physical phenomenon is also common among pupils. Thus, the idea of light scattering or diffuse reflection is usually not familiar to learners at the beginning of the optics class. Children and adolescents find it difficult to believe that objects that do not shine by themselves can emit light when illuminated by a light source. Earth's moon is an exception. Here, the idea that the Moon is illuminated by the Sun and that light rays from the surface of the Moon are scattered in different directions appears plausible to the learners (Haagen-Schützenhöfer & Hopf, 2018). Experience has shown that it is, therefore, advisable to use exactly this phenomenon to introduce diffuse reflection.

Diffuse reflection of light is one of the basic ideas in optics lessons because it also provides a physical explanation of the visual process. Reflection on smooth or reflecting surfaces is generally understood from everyday life and is also referred to in physics as regular reflection. The angle of incidence and the angle of reflection are the same (Fig. 22.12). In contrast, diffuse reflection or light scattering describes the phenomenon that light rays on non-smooth surfaces are random, i.e. reflected in different directions (Tipler & Mosca, 2008).

As previously mentioned, diffuse reflection forms the basis for understanding the physical part of the visual process. An elementary explanation of vision, which is common in secondary school, is as follows: the light from the light source reaches objects in this area, it is diffusely reflected or scattered on the surface of these objects and then catches the eye. The light rays on the cornea and the eye-lens are refracted, and an image of the object is created on the retina.

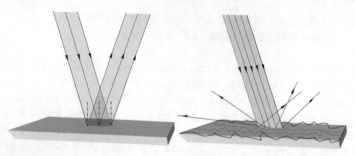

Fig. 22.12 Left: Specular reflection of light at a reflecting surface, parallel light rays of a light cone are reflected at a surface at the same angle to the perpendicular (dashed lines). Right: diffuse reflection at a non-mirror-like surface, parallel light rays of a light cone are reflected at different angles due to the irregular surface

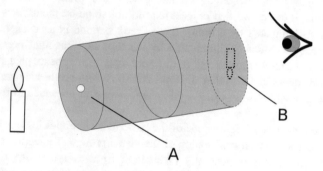

Fig. 22.13 Principle of a pinhole camera. **A**: Pinhole, **B**: Screen

When pupils are asked about the visual process, they rarely establish a relationship between the object and the eye. Many learners do not think it necessary that the light from the object catches the eye to be visible to us (Haagen-Schützenhöfer & Hopf, 2018).

The starting point of the learning process in physics is largely determined by prevailing preschool perceptions. A successful transfer of knowledge in physics classes and geometrical optics, in particular, can only take place if teachers are informed about the pupil's ideas and take them into account in their lesson planning.

22.3.3 Image Scale and Geometrical Theorems as the Basis of Geometrical Optics

Mathematics and physics are closely linked not only in science but also in lessons at school. In the context of geometric optics, this link can be demonstrated, for example, by the derivation of the image scale. Here, the relevance of the content from math lessons for comprehending the connections in physics lessons is not only directly visible to teachers but also learners.

The image scale A is the ratio of the image size I to the object size O or the quotient of the image width i and the object width o.

$$A = \frac{I}{O} = \frac{i}{o}.$$

The image scale can be developed in the classroom, for example, using image formation with the camera obscura or the so-called pinhole camera. The camera obscura is an apparatus for generating images and has a simple structure. In a light-tight room, a narrow hole is created through which the light from objects can strike the back wall of the room and create an image of these objects.

The image created in this way is reversed and upside down. If the back wall is transparent, the image can also be viewed from the outside. If a small box is used for the image generation, it is called a pinhole camera because the hole is mostly pierced with a pin.

The camera obscura and the pinhole camera are created according to the following principle: the light rays emanating from the object are limited by the pinhole to narrow light beams with a small opening angle. This means that only the light rays transmitted represent the direct connection from the object to the point. Rays from the top of the object coincide with the bottom of the figure and vice versa. The light intensity can be regulated by the size of the pinhole; the larger the hole, the more light-intensive the image. Although, the image loses its sharpness, because the smaller the pinhole, the sharper the image.

The pinhole camera can be used in school lessons for various reasons. This camera is critical in the history of science, as it is the forerunner of the photo camera, the solar projection (for observing solar eclipses and sunspots) or even the photocopying machine. From an educational point of view, the discussion or construction of the pinhole camera in physics lessons is also recommended, because the formation of images on the pinhole camera can be explained by relatively simple observations with the help of light rays and can, thus, explain image formation on optical components such as lenses. By encouraging students to build their own pinhole cameras, teachers can expand students' motivation and self-activity. Also, from the point of view of mathematics teaching, it is highly recommended to mainly treat the pinhole camera in physics lessons experimentally. In this way, teachers can create an action-oriented basis for the derivation or application of the intercept theorem.

The image scale (and thus also the second set of rays) can be derived in secondary school lessons using the drawing below (Fig. 22.14; extended sketch of the image formation in a pinhole camera).

In this sketch, the rays in a pinhole camera are shown. The observed object with the size O is on the left with the edge points A and E and the central point F. The pinhole is in point Z, the distance between the object and the hole $\overline{FZ}$ is called object distance o. The picture of the object is on the right, with the size I, going from point D to point C with the central point H. The distance i between the hole and the picture is called the image distance.

A prerequisite for the derivation is that the theorems of similarity for triangles have already been resolved. According to one of the similarity theorems, triangles are similar to each other if they match at two angles (and due to the sum of the angles in a triangle, they also match in three angles). In Fig. 22.15, the triangles $\triangle AFZ$ and $\triangle CHZ$ are similar because their angles match. Here $\alpha = \alpha\prime$ because they are vertical angles and $\beta = \beta\prime$ because they are alternate angles.

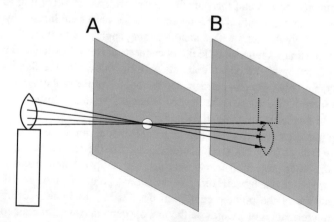

Fig. 22.14 The principle of image formation on a pinhole camera. **A**: Pinhole, **B**: Screen

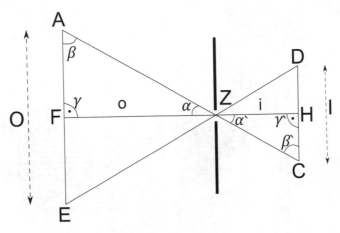

Fig. 22.15 Geometric considerations of image formation on a pinhole camera. O: object size, i: image size, o: object distance, i: image distance

The corresponding side lengths are in the same ratio in similar figures. The following ratio, therefore, results for corresponding routes: $\frac{\overline{CH}}{\overline{AF}} = \frac{\overline{HZ}}{\overline{FZ}}$. Accordingly, the following applies: $\frac{I/2}{O/2} = \frac{i}{o}$ and also: $\frac{I}{O} = \frac{i}{o}$. As previously explained, this ratio is referred to as the image scale.

From this example, it becomes clear that the content of geometrical optics, if a deeper understanding is sought, is accompanied by mathematical teaching. As a rule, content from geometrical optics is taught in the lower classes of lower secondary school, but the content of the mathematics lessons necessary for deduction is not yet discussed at this time. In this case, a spiral curriculum could solve the problem. A spiral curriculum involves teaching the same topics several times and in the later grades at a higher level. However, this procedure is based on a detailed didactic reduction for the lower classes, as this reduced presentation can be used at a later date.

22.3.4 Thin Lens Equation and Experiments in Physics Education

Images of objects can also be created with the help of optical components such as thin lenses or spherical mirrors. This type of image generation is used, for example, in the case of glasses; surveillance mirrors at the cash register of a supermarket; or forehead mirrors in ear, nose and throat medicine. The relationship between

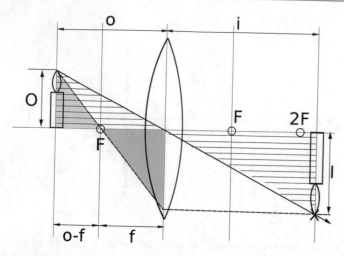

Fig. 22.16 Derivation of the image scale. O: object size, I: image size, o: object distance, i: image distance, f: focal length, F: focal point. (According to Meyer, 2003, p. 330)

the focal length f, the image distance b and the object distance g describes the so-called thin lens equation.[3] This is:

$$\frac{1}{o} + \frac{1}{i} = \frac{1}{f}.$$

If the image scale A (defined as the ratio of image and object size $A = \frac{I}{O}$ or as the ratio of image and object distance $\frac{i}{o}$) is also taken into account, the lens equation can be transformed to determine, for example, image and object distances at a given magnification:

$$i = (A + 1) \cdot f \text{ or } o = \left(\frac{1}{A} + 1\right) \cdot f.$$

In school at the lower secondary level, this equation is derived analogously to the image scale using the similarity theorems or the intercept theorems (see Fig. 22.16). Figure 22.16 shows how an image of size I of an object of size O (here, a candle) can be obtained with the help of a lens. The image is constructed

---

[3] See chapter "Educational use of Ludwig's methodology using the example of the lens equation".

with two light rays, namely the chief ray and the focal ray. These rays of light and other auxiliary lines form the basis of the geometrical considerations. The distance o from the center of the object to the center of the lens is the object distance. Accordingly, i is the image distance. F is the focal point of the lens and f is the focal length of the lens.

From the triangles with a light gray background, the intercept theorem shows that.

$\frac{I}{O} = \frac{i}{o}$, that is to say the image scale A already explained. According to the intercept theorem, the following is deduced from the triangles marked in dark gray:

$$\frac{O}{I} = \frac{o - f}{f}.$$

Replacing $\frac{O}{I}$ with $\frac{o}{i}$ and transforming accordingly results in the thin lens equation:

$$\frac{1}{o} + \frac{1}{i} = \frac{1}{f}.$$

In connection with the thin lens equation, the focal length of a converging lens is often experimentally determined in optics lessons. The experimental determination is based on the fact that the thin lens equation enables the focal length to be calculated by appropriate transformations if the object distance o and the image distance i are known.

$$f = \frac{o \cdot i}{o + i}.$$

To experimentally determine the focal length, thin converging lenses with different unknown focal lengths are given to the learners. An arrangement, as visible in Fig. 22.17, is set up on an optical bench, and by repeatedly moving the lens or the optical screen until there is a sharp image on the screen, pairs of values for the object distance and image distance are recorded. Based on the imaging equation, the focal lengths for a converging lens are calculated with the help of the corresponding pairs of values. Then, an average value is formed for the experimentally determined focal length. The comparison with the focal length specification, provided by the manufacturer, then gives us an image of how good the measurements are or how well the focal length can be determined by the described method.

Due to the outstanding role of the experiment in physics lessons, students should acquire so-called experimental competence in the classroom and constantly develop it further. According to Schreiber et al. (2009), the sub-competences of the experimental competence include the formation of hypotheses, the planning and execution of experiments, and the analysis of experiments.

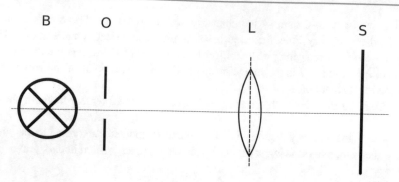

Fig. 22.17 Schematic experimental setup for the experimental determination of the focal length of a converging lens. Components: B: light source, O: object mount with the object to be illuminated (e.g. a reversal film), L: converging lens, S: screen

Accordingly, experimental competence is much more than just carrying out an experiment. According to Schreiber et al., (2009), the sub-component "carrying out an experiment" alone comprises several fields of activity, such as putting together the test equipment, arranging the experimental set-up, recording the measured values and documenting the results. In addition, there are the sub-steps of planning (formulating the questions or expectations of the experiment and forming hypotheses) as well as evaluation (preparation and processing of the data and interpretation of the results).

Schreiber et al. (2009) note that hypothesis-based experimentation in class and scientific experimentation in physics research should be clearly distinguished from one another despite their similar structures. Muckenfuß (2010) also expresses his concerns about this common confusion and justifies the difference between hypothesis-based experimentation in class and science, as the classroom is usually a condensed version of scientific hypothesis-based experimentation acts because experiments are carefully prepared or developed in advance by teaching material manufacturers and teachers.[4]

The experimental determination of the focal length of a converging lens with the help of the thin lens equation shows, for example, how experiments are arranged in physics lessons. Experiments are an essential feature of the natural sciences and are, therefore, also of central importance in science lessons (Kircher,

---

[4] See also Volume 1, Chapter „ Development of Knowledge in Mathematics and Physics Education " (Dilling, Stricker, Tran, Vu, 2020).

2015). Experiments in the classroom can fulfill various functions. According to Girwidz (2015) they can be used to arouse motivation and interest, test pupils' ideas, provide food for thought, demonstrate a phenomenon or build up a physical idea, verify a theoretically derived law, etc. (see Vol. 1 'Development of Knowledge', pp. 326). Experiments can also occur in different forms within a class (teacher or student experiments) and be used in different teaching phases. The determination of the focal length of a converging lens in the form described above is an experiment that is most often used by students in the elaboration phase for the quantitative verification of a theoretically developed law. Depending on the course of the lesson, a teacher may use the described experiment as a teaching or demonstration attempt. The experiments in class, therefore, cannot be clearly assigned to a specific category or function; these can only serve a specific purpose in connection with the course of the class. One experiment can combine different functions and, therefore, also be used for different purposes.

The interplay of geometric, algebraic, and scientific (more precisely physical) views on the same issue is well illustrated by the topic of lens equation. If one of the views is missing, the insights cannot be completely obtained. Such an interdisciplinary approach, which is common in science, also proves to be profitable in the classroom, since this approach enables a representation of the methods of work in research that is as close to reality as possible.

22.4 Conclusion and Outlook

This chapter demonstrated that many relations exist between the topics of straight lines, similarity and intercept theorems from mathematics and the topics of light rays, scattering of light, images and lenses from physics. In school, the knowledge gained in a mathematics class on plane geometry forms an explanatory basis for many physical phenomena from the field of optics. Through this authentic application of mathematical knowledge, students can perceive the connection between mathematics and reality in a special way. If one examines the historical development of geometry and, in particular, the field around the intercept theorem, one can recognize that these were also induced from a practical (physical) problem. This simple example succinctly shows that mathematics and physics have many connections.

The subject-specific as well as subject-didactic-specific approaches presented in this chapter can equally enrich the teaching of mathematics as well as physics or even integrated STEM education. Geometric optics offers an accessible application of the intercept theorem and, at the same time, is the explanatory basis of many physical phenomena. To satisfy both the mathematical and the physical perspective, teachers of each subject should be sufficiently informed about the background of the other subject. This chapter can prompt such an examination.

References

Dilling, F., & Vogler, A. (2020b, in print). Ein mathematisches Zeichengerät (nach) entwickeln—eine Fallstudie zum Pantographen. In F. Dilling & F. Pielsticker (Eds.), *Mathematische Lehr-Lernprozesse im Kontext digitaler Medien* (pp. 103–126). Springer Spektrum.

Dilling, F., Marx, B., Pielsticker, F., Vogler, A., & Witzke, I. (2020, in print). *Praxisbuch 3D-Druck im Mathematikunterricht. Einführung und Unterrichtsentwürfe für die Sekundarstufe I und II.* Waxmann.

Dilling, F., Stricker, I., Tran, N. C., & Vu, D. P. (2020). Development of knowledge in mathematics and physics education. In S. F. Kraus & E. Krause (Eds.), Comparison of mathematics and physics education I : Theoretical foundations for interdisciplinary collaboration (pp. 299–344). Springer Fachmedien Wiesbaden. https://doi.org/10.1007/978-3-658-29880-7_13

Duit, R. (2004). PIKO-BRIEF NR. 1, MAI 2004. Schülervorstellungen und Lernen von Physik. http://www.idn.uni-bremen.de/schuelervorstellungen/material/Piko-Brief_Schuelervor.pdf, Accessed 5. Apr. 2020.

Gericke, H. (1984). *Mathematik in Antike und Orient.* Springer.

Girwidz, R. (2015). Medien im Physikunterricht. In E. Kircher, R. Girwidz, & P. Haußler (Eds.), *Physikdidaktik.* Springer.

Haagen-Schützenhöfer, C., & Hopf, M. (2018). Schülervorstellungen zur geometrischen Optik. In H. Schecker, T. Wilhelm, M. Hopf, & R. Duit (Eds.), *Schülervorstellungen und Physikunterricht. Ein Lehrbuch für Studium, Referendariat und Unterrichtspraxis* (pp. 89–114). Springer.

Helmerich, M. A., & Lengnink, K. (2016). *Einführung Mathematik Primarstufe—Geometrie.* Springer Spektrum.

Hölzl, R., et al. (2018). Ähnlichkeit. In H.-G. Weigand (Ed.), *Didaktik der Geometrie für die Sekundarstufe I* (pp. 203–226). Springer Spektrum.

Jung, W. (1979). *Aufsätze zur Didaktik der Physik und Wissenschaftstheorie.* Diesterweg.

Kircher, E. (2015). Über die Natur der Naturwissenschaften lernen. In E. Kircher, R. Girwidz, & P. Häußler (Eds.), *Physikdidaktik.* Springer.

Meyer, L., & Schmidt, G.-D. (2003). *Duden. Basiswissen Schule.* Dudenverlag.

Muckenfuß, H. (2010). *Experimentieren und Versuche machen: Erkenntnistheoretische Aspekte der Sachbegegnung im naturwissenschaftlichen Unterricht* [Veröffentlichung im Rahmen des 23. Karlsruher Didaktik-Workshops].

Schreiber, N., Theyßen, H., & Schecker, H. (2009). Experimentelle Kompetenz messen?! *PhyDid a, Physik Und Didaktik in Schule Und Hochschule, 8*(3), 92–101.

Struve, H. (1990). *Grundlagen einer Geometriedidaktik.* Bibliographisches Institut.

Tipler, P. A., & Mosca, G. (2008). *Physics for scientists and engineers. Standard* (6th ed.). W.H. Freeman.

Tran, N. C., Chu, C. T., Holten, K., & Bernshausen, H. (2020). Models and modeling. In S. F. Kraus & E. Krause (Eds.), Comparison of mathematics and physics education I : Theoretical foundations for interdisciplinary collaboration (pp. 257–298). Springer Fachmedien Wiesbaden. https://doi.org/10.1007/978-3-658-29880-7_12

Vu, D. P., Nguyen, V. B., Kraus, S. F., & Holten, K. (2020). Individual concepts in physics and mathematics education. In S. F. Kraus & E. Krause (eds.), Comparison of mathematics and physics education I : Theoretical foundations for interdisciplinary collaboration (pp. 215–256). Springer Fachmedien Wiesbaden. https://doi.org/10.1007/978-3-658-29880-7_11

Weinmann, K. F. (1980). *Die Natur des Lichts. Einbeziehung eines physikgeschichtlichen Themas in den Physikunterricht.* Wissenschaftliche Buchgesellschaft.

Lesson Plan: Straight Lines and Conic Sections

Le Tuan Anh

Lesson Title: Applying conics to physics and the real world
Abstract: This teaching sequence is designed to apply the concepts and properties of ellip-ses, hyperbolas, and parabolas. For these concepts and properties, which are known from a purely mathematical point of view, the students should inde-pendently search for different real applications. The task formats range from an open-ended search for buildings to concrete physical tasks. Methodically, the focus is on joint development and documentation in groups.

Type of school / Grade	High School / Grade 10
Prerequisites	– Knowledge of concepts and properties of ellipses, hyperbolas, and parabolas
Number of periods	2
Objectives	Mathematics – Students can consolidate the concepts and properties of conics Physics – Students understand the applications of the conics in physics and in reality

L. T. Anh (✉)
Faculty of Mathematics and Informatics,
Hanoi National University of Education, Hanoi, Vietnam
e-mail: anhlt@hnue.edu.vn

© The Author(s), under exclusive license to Springer Fachmedien Wiesbaden GmbH, part of Springer Nature 2022
F. Dilling and S. F. Kraus (eds.), *Comparison of Mathematics and Physics Education II*, MINTUS – Beiträge zur mathematisch-naturwissenschaftlichen Bildung, https://doi.org/10.1007/978-3-658-36415-1_23

23.1 Methodical Commentary

The students have learned the concepts and properties of conics including ellipses, hyperbolas, and parabolas. These lessons concentrate on applying these shapes in physics and finding them in reality. Conics have many applications, including buildings and architecture. Through these tasks, the students learn some applications of conics, which can be divided into the following areas: buildings, architecture, and real life; astronomy and cosmology.

The aim of the lesson is to allow students to find conic sections independently in a variety of contexts. To enable a consistent evaluation for the presentation of the results, rubrics are set up in the first hour. They include cooperation in the groups, the correctness of the collected information, and the content and preparation of the presentations.

23.2   Lesson Plan

Nr. & Time	Stage	Learning activities	Forms of Interaction	Materials/ Resources	Methodological remarks
# 1 30 min	Approaching the task	*Approaching the task:* – Teacher divides the class into several groups; – Each group receives the task of discovering applications of conics in physics and reality; – Teacher builds the rubrics to assess the groups' results	Teacher presentation:/ Whole class discussion	Textbooks, Internet	The rubrics should assess different aspects of each group, including cooperation, the correctness of the information and contents and the appearance of the presentation
# 2 outside the class (around one week)	Working on the project	*Group working on the project:* A student in each group receives his/her task and does the task Each group discusses the task, finds information, writes a report, and prepares a presentation	Individual work/ Group work/ Project-based learning	Textbooks, Internet	The students in each group can meet face-to-face or discuss using other devices such as email, Facebook, telephones, and cell phones
# 3 60 min	Presentation	*Presenting the results:* Each group presents its results; the other group and the teacher give their comments to improve the results	Group work/ Whole class discussion	The product of groups, the rubrics to assess	Each group presentation can be carried out according to the following process: • The group presents its results; • The other groups and the teacher give their comments on the performance • The other groups and the teacher assess the presenting group according to the rubrics • After the presentations, the teacher sums up, gives the overall assessment

Tasks are given to the groups:
Task 1

(i) Find the appearances of conics in real life.
(ii) Find the applications of conics in buildings, architecture, etc. Explain some benefits of conic shapes in buildings and architecture.

Task 2

(i)   Finding the initial velocity of a spaceship launched from the Earth and its orbit.
(ii) Discovering the orbits of the planets in the solar system.
      Discovering the eccentricity of the orbits of the planets in the solar system. Which orbits look like circles? Why?
(iii) Discovering the orbit of the moon around the Earth.

Task 3

Discovering the optical properties of light rays and applications.

Task 4

Solving some application problems related to conics:

Problem 1 (Whispering Gallery). If an elliptical whispering gallery is 25 feet wide and 80 feet long. What is the distance between two people in the room so that they can hear each other whispering? (Lippman & Rasmussen, 2020, p. 596).

Problem 2 (Planetary Orbits). The orbits of planets around the sun are nearly elliptical with the sun as their focus. The perihelion is a planet's shortest distance from the sun and the aphelion is its longest. The length of the major axis is the sum of the perihelion and the aphelion. What is an equation for Earth's orbit, given that Earth's perihelion is 91.40 million miles, and its aphelion is 94.51 million miles? (Lippman & Rasmussen, 2020, p. 596).

Problem 1 (the LORAN) Two stations A and B are 100 km apart, and they send a simultaneous radio signal to a ship. The signal from one arrives 0.0002 s after the signal from the other. Suppose that the signal goes 300,000 km per second, write an equation for a hyperbola on which the ship is located if A and B are the foci of the hyperbola. (Lippman & Rasmussen, 2020, p. 615).

Some intended results:
Task 1

(i) The appearances of conics in real life:
- Tilt a glass of water and the surface of the liquid acquires an elliptical outline (Fig. 23.1).
- The shadow of a circle on a plane is an ellipse (Fig. 23.2).
- A stream of water from a fountain has a parabolic shape (Fig. 23.3).
- A shadow of a table lamp on a wall forms a hyperbola (Fig. 23.4).

Fig. 23.1   A tilted glass of water

Fig. 23.2   The shadow of a circle

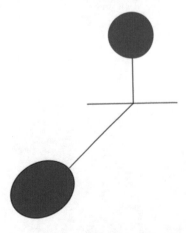

Fig. 23.3 A water fountain. (License: CC0, https://creativecommons.org/publicdomain/zero/1.0/legalcode)

Fig. 23.4 The shadow of a table lamp on the wall

Some other examples are the shadows of a ball, elliptical shapes of billiards tables, a fall of parabolic track of roller coaster, the orbit of an object thrown into the air, and the elliptical orbit of electrons moving around the nucleus.

(ii) Properties of conics are applied in building, architecture:

- The parabolic gate of Hanoi University of Science and Technology (Fig. 23.5).

Some other examples are the Eiffel Tower, Whispering galleries at St Paul's Cathedral in London, The Parabola in London in 1962, The Golden Gate Bridge

Fig. 23.5 The gate of Hanoi University of Science and Technology. (Source: wikimapia. org, license: CC BY-SA 3.0, https://creativecommons.org/licenses/by-sa/3.0/legalcode)

Fig. 23.6 Tycho Brahe Planetarium in Copenhagen. (Author: Flemming Rasmussen, license: CC BY-SA 4.0, https://creativecommons.org/licenses/by-sa/4.0/legalcode)

in San Francisco in California, an ellipse-shaped park in front of the White House in Washington D.C., hyperbolic parabolic Dulles Airport, hyperbolic McDonnell Planetarium, parabolic Rialto Bridge (Venice, Italy), parabolic Arc De Triomphe in Paris, France, elliptic SA1 Swansea Waterfront (UK), hyperbolic cooling towers of a nuclear reactor, hyperbolic gears of a transmission, parabolic mirrors in solar ovens (e.g. Fig. 23.7).

Using conic shapes in buildings, architecture, etc. has many advantages (Fig. 23.6). Semielliptical arches have been used for centuries to build bridges like the Skerton Bridge of England. They allow for longer spans between

Fig. 23.7 The Kobe Port Tower. (Author: 663highland, license: CC BY 2.5, https://creativecommons.org/licenses/by/2.5/legalcode)

supports. The National Statuary Hall in Washington, D. C. is often called a whispering chamber because the ellipse-shaped room allows two people standing at the foci of the ellipse to hear each other whispering although they are far apart. Before the GSP, the LORAN (Long Range Navigation) system was used to find a ship's location by using two pairs of radio stations and determining the intersection of two hyperbolas with these stations as their foci.

Task 2

(i) A spaceship launched from the Earth can fly with the orbit of a part of a hyperbola, a part of a parabola, an ellipse, and a circle. The following table shows the relationship between the initial velocity of the spaceship and its orbits.

The initial velocity v_0 of a spaceship	The form of the orbits of a spaceship
$v_0 > 11.2$ km/s	A part of a hyperbola
$v_0 = 11.2$ km/s	A part of a parabola
7.9 km/s $< v_0 < 11.2$ km/h	An ellipse
$v_0 = 7.9$ km/s	A circle

In addition, scientists also calculate general cosmic velocities, that is, the speed of objects moving relative to other bodies under the influence of mutual gravity. They do this, for example, to launch a spacecraft. For the escape cylinder to return to the Earth from the Moon, it must give the spacecraft an initial speed of 2.38 km/s.

- The orbits of the planets around the Sun are ellipses, and the Sun is one of their foci.
- The eccentricity of the orbits of the planets in the solar system (Fig. 23.8):

Venus:	$e \approx 0{,}0068$	Mars:	$e \approx 0{,}0934$
Jupiter:	$e \approx 0{,}0484$	the Earth:	$e \approx 0{,}0167$
Mercury:	$e \approx 0{,}2056$	Neptune:	$e \approx 0{,}0082$
Saturn:	$e \approx 0{,}0543$	Uranus:	$e \approx 0{,}0460$

The orbits of Venus, the Earth, and Neptune look like circles because of their eccentricities.

(iii) The moon moves around the Earth with the orbit of an ellipse with eccentricity $e \approx 0.0549$, and the Earth is one of its foci.

Fig. 23.8 The solar system. (License: public domain)

Task 3

- *The optical properties of light rays*

Light emitted from one focus that hits any point on the ellipse passes through the other focus. The light rays directed toward the focus of a hyperbola are reflected by a hyperbolic mirror to the other focus of the hyperbola. (Fig. 23.9).

Light originating at the focus of a parabola is reflected parallel to the axis of the parabola, and light arriving parallel to the axis is directed toward the focus (Fig. 23.10).

- Applications of the optical properties of light:

The following are some applications of the optical properties of light.

If a bulb is placed at the focus of a parabolic mirror, the light rays reflect off the mirror parallel to each other, making a focused beam of light (Fig. 23.11).

A radio telescope typically has three components: antennas, a receiver and amplifier, and a recorder. Most antennas are parabolic dishes that reflect the radio

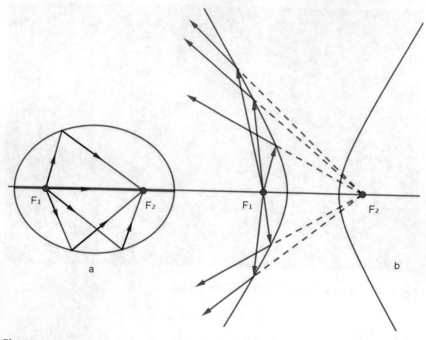

Fig. 23.9 The property of the elliptical and hyperbolic mirrors

Fig. 23.10 The property of
a parabolic mirror

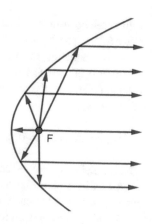

Fig. 23.11   A car headlight

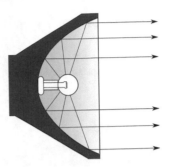

waves to a receiver, in the same way as a curved mirror focuses visible light to a point (Fig. 23.12).

The reflecting property of an ellipse is used in pulverizing kidney stones.

Fig. 23.12   A radio telescope. (Author: Lapinov, license: CC BY 4.0, https://creativecom-mons.org/licenses/by/4.0/legalcode)

Reference

Lippman, D., & Rasmussen, M. (2020). *Precalculus: An investigation of functions*: https://www.opentextbookstore.com/precalc/.

Printed in the United States
by Baker & Taylor Publisher Services